Leading G

Early Praise for Leading Global Innovation

"Karina Jensen has managed not only to give an up-to-date overview of the crucial aspects of the innovation process, but she has also given quite some attention to the multicultural environment as an incubator for innovation. A must-read for those interested in how diversity and inclusion can help your organization become more innovative."
—Fons Trompenaars, Author of best-selling *Riding the Whirlwind: How to Create a Culture of Innovation*

"Based on case studies of multinationals, this book provides practical insights on driving innovation in a culturally diverse global environment. Timely book for trying times."
- Vijay Govindarajan Coxe Distinguished Professor, *Tuck School of Business at Dartmouth and Marvin Bower Fellow, Harvard Business School*

"A thoroughly researched and insightful exposé of the challenges and means to drive a successful global innovation culture, agenda, and in-market success. Well done!!"
—Bryan Semkuley, Vice President, *Global Innovation, Kimberly-Clark Professional*

"A fascinating and welcome look at the merger of multicultural understanding and global innovation. Based on extensive research, this book captures the challenges and solutions for facilitating multicultural collaboration in the crucial realm of innovation."
—Joyce Osland, Lucas Endowed Professor of Global Leadership and Executive Director, *Global Leadership Advancement Center, San Jose State University*

"In our global economy, one of the key success factors in sustaining global innovation requires organizations to provide a creative space and open environment that supports multicultural team collaboration. Karina Jensen's book provides a framework and guideline supporting organizational goals and ambitions."
—Sheryl Chamberlain, Vice President, *Group Strategic Initiatives and Partnerships, Capgemini*

"If you are looking to transform your firm to become a truly global innovator this book is a must-read. The insights are extremely practical and useful and will help any organization looking to innovate."
—Brian Lawley, *CEO and Founder, 280 Group*

"Innovation in the 21st century is being played out across an increasingly wide geographical stage. This book offers an excellent lens through which to explore some of these shifts and in particular helps us understand some of the strategic management challenges involved."
—John Bessant, Chair and Professor of *Innovation and Entrepreneurship, University of Exeter*

Karina R. Jensen

Leading Global Innovation

Facilitating Multicultural Collaboration
and International Market Success

Karina R. Jensen
Paris, France

ISBN 978-3-319-53504-3   ISBN 978-3-319-53505-0 (eBook)
DOI 10.1007/978-3-319-53505-0

Library of Congress Control Number: 2017943379

© The Editor(s) (if applicable) and The Author(s) 2017
This work is subject to copyright. All rights are solely and exclusively licensed by the Publisher, whether the whole or part of the material is concerned, specifically the rights of translation, reprinting, reuse of illustrations, recitation, broadcasting, reproduction on microfilms or in any other physical way, and transmission or information storage and retrieval, electronic adaptation, computer software, or by similar or dissimilar methodology now known or hereafter developed.
The use of general descriptive names, registered names, trademarks, service marks, etc. in this publication does not imply, even in the absence of a specific statement, that such names are exempt from the relevant protective laws and regulations and therefore free for general use.
The publisher, the authors and the editors are safe to assume that the advice and information in this book are believed to be true and accurate at the date of publication. Neither the publisher nor the authors or the editors give a warranty, express or implied, with respect to the material contained herein or for any errors or omissions that may have been made. The publisher remains neutral with regard to jurisdictional claims in published maps and institutional affiliations.

Cover illustration: artvea/ DigitalVision Vectors/Getty

This Palgrave Macmillan imprint is published by Springer Nature
The registered company is Springer International Publishing AG
The registered company address is: Gewerbestrasse 11, 6330 Cham, Switzerland

*To my parents
for inspiring me to see the world*

Preface

Enjoy the Multicultural Learning Journey

Traveling around the world, from Paris to Kathmandu to Bangalore to San Francisco, it has been interesting to witness the collaborative needs from very poor to wealthy communities who need to connect with a global network, solving problems as diverse as education access and professional development to customer solutions and international market strategies.

Living in an increasingly interconnected and interdependent world provides the opportunity to learn and benefit from a wealth of culturally diverse perspectives and practices. Yet this potential is not fully optimized within our communities and organizations. The challenges of time, trust, distance, and communication contribute to failed collaborations. The outcome? A lack of innovation, efficiency, performance, and results.

A global and dynamic business environment is quickly evolving with changing needs where innovation is a constant. Leaders are facing the challenges of navigating a multicultural and digitally connected work environment where collaboration is the norm. The acceleration of competition from diverse international markets requires a responsive organization. New innovations demand consideration of their cultural and local fit for consumers in mature and emerging markets. The outcome is an increased focus on multicultural collaboration in order to ensure organizational performance and international market success.

This book argues that multicultural collaboration is central to global innovation through the integration of new ideas and diverse perspectives from around the world. The focus therefore becomes how to optimize multicultural collaboration and knowledge-sharing for global innovation success. Culturally diverse views and knowledge bring valuable insights to local challenges and opportunities that are critical to the creation of international solutions. It's important to link local voices to global actions in order to create solutions that truly speak to the culturally diverse markets that are served by organizations worldwide. This requires new competencies for leaders who will need to serve as facilitators and orchestrators of multicultural innovation and collaboration.

Then how do we create leaders, teams, and organizations capable of collaborating in an increasingly multicultural and networked environment? It is a question that has served as great fascination throughout my career, from Stockholm to San Francisco to Paris, as I gained cross-cultural perspectives in my international marketing and sales roles. These were years of great insight and rich in experience where I had the opportunity to work for multinational firms in launching new products and services to international markets in Asia-Pacific, Europe-Middle East-Africa, North and South America. This led to the creation of a consulting firm, Global Minds Network, focused on global innovation and collaboration solutions. My corporate and consulting experiences with multinational high-tech firms in Silicon Valley provided me with an up-close view of recurrent issues: the challenges of cross-cultural team collaboration when creating and delivering new concepts to markets around the world.

This global innovation challenge turned into an idea for a PhD thesis in order to provide the time and opportunity to fully investigate key issues and ultimately discover new solutions. And so, I packed my bags and left the fast-paced and entrepreneurial high-tech region of the San Francisco Bay Area for a PhD program in the cultural and creative city of Paris. The one-year work sabbatical transformed into a four-year learning journey and world tour in order to find the answer to the research question: How can multinational firms facilitate cross-cultural team collaboration in order to accelerate global innovation management capabilities?

The learning adventure and research project employed empirical qualitative research through on-site meetings and phone interviews with leaders at 40 multinational firms in 14+ countries who were leading and facilitating global innovation and launch projects involving cross-cultural teams while based in Asia, Europe, and North America. At the end of the PhD journey in 2012, a framework for global collaboration emerged thanks to new discoveries and findings. Since then, new frameworks and models have evolved with an extended global study as well as a regional study in Asia from 2012–15.

Today in my work as a consultant, researcher, and educator, I enjoy the opportunity to explore new challenges and solutions for leading global innovation, facilitating change, and collaborating with cross-cultural teams.

The learning journey continues and the quest has become the creation of multicultural innovation and collaboration models and practices that improve organizational performance and international market success for leaders, teams, and organizations.

In this book, you will discover the Multicultural Innovation Framework and its key collaboration drivers – Vision, Dialogue, and Space. It takes a holistic view that examines the role of leadership in facilitating global collaboration and innovation success. For each section, you will find highlights including critical discoveries and practices from the studies, as well as recent case examples from multinational firms. A visual model is presented at the start of each section to guide you on this international learning journey, inviting you to explore and discover the keys to leading multicultural innovation and collaboration.

Bon voyage
Karina R. Jensen

Acknowledgments

I would like to extend my thanks and great appreciation to the numerous leaders and study participants whose generous time, insights, and collective wisdom have created the road map for my research on multicultural innovation and collaboration.

A special thank you goes to my key contributors for providing valuable insights and authoring some of the case studies featured in this book: Hans-Juergen August, Vice President of Innovation and Quality Management at Siemens Convergence Creators, who provided a valuable case on effective global innovation and collaboration practices; Thomas Arend for an insightful account of his impressive international roles at Google, Airbnb, Twitter, and beyond; Balamurugan Kannan, General Manager of BFSI Global Delivery and Saksham Khandelwal, Member of Thought Leadership Charter at Wipro for a truly innovative case for global and local collaboration practices; and Kurt Ward, Senior Design Director of Philips for sharing leading practices in co-creation and design thinking.

A grand merci to my academic colleagues who offered support and advice for this book project: Nigel Holden, Dana Minbaeva, Fiona Moore, Jean-Paul Lemaire, and Dominique Xardel. Thank you to my colleagues at NEOMA Business School for your support of this book project: Chris Worley, Annabel-Mauve Adjognon, Svetlana Serdyukov, and Lisa Thomas. My sincere appreciation to Bryan Semkuley, Hans-Juergen August, Adam Travis, Clynton Taylor, Sheryl Chamberlain, Fons Trompenaars, Joyce Osland, and Brian Lawley for your invaluable feedback in reviewing this book.

I would also like to thank Liz Barlow at Palgrave Macmillan for your great interest and support in making this book project a reality. An extended thank you to Lucy Kidwell, the editorial, design, and production teams at Palgrave Macmillan, for skillfully managing the execution process.

Finally, I would like to extend my deepest appreciation to my family and good friends who offered endless support and encouragement during this journey.

Contents

1 Introduction: Achieving Innovation in a Global and Dynamic Environment — 1
The Evolving and Dynamic Global Business Environment — 2
Accelerating Global Innovation Through Multicultural Collaboration — 5
Bibliography — 10

Part I Vision: Global Leadership and Strategic Co-Creation

2 Leading Global Innovation and Collaboration — 13
Demands of a Multicultural Organization — 14
The Global Leadership Development Journey — 16
Optimizing Multicultural Team Collaboration — 20
New Leadership Model for Multicultural Innovation and Collaboration — 21
Global Leaders as Facilitators and Orchestrators — 30
Bibliography — 31

3 From Global Strategy to Strategic Co-creation — 33
Innovation Strategies for Mature and Emerging Markets — 34
Internationalizing Radical and Incremental Innovation — 37
Optimizing Creativity through Cultural Diversity — 39
The Challenges of Global Strategic Planning — 40
Local Roles within an Interdependent Organizational Network — 44

From Strategic Planning to Strategic Co-creation	48
Bibliography	52

Part II Dialogue: Nurturing Knowledge-Sharing and Learning

4 Communicating in a Multicultural and Networked World — 57
The Role of Culture in Global Innovation	58
The Dynamics of Cross-Cultural Communication	58
Language and Project Communication	60
Cultural Differences in Knowledge-Sharing	61
Knowledge-Sharing and Multicultural Innovation	62
Learning from Eastern and Western Perspectives	70
Bibliography	74

5 Listening to Local Market Voices — 77
Facilitating Multicultural Team Collaboration, from Concept to Market	78
Building Trust Across Cultures and Functions	82
Motivation and Multicultural Team Collaboration	85
Collective Wisdom through Collaboration	90
Bibliography	94

Part III Space: Creating an Environment for Inclusive Innovation

6 Developing a Global Innovation Culture — 97
Nurturing an Inclusive Environment for Multicultural Organizations	98
The Drivers of a Global Innovation Culture	99
Optimizing a Global Innovation Culture Through Organizational Routines	105
Bibliography	111

7 Creating an Inclusive Innovation Climate — 113
The Global Innovation Cycle and Project Process	114
Multicultural Project Collaboration, from Planning to Execution	115
Communication in Complex, Multicultural Team Contexts	119
Development of an Inclusive Innovation Climate	120

Enabling Local Connections to Customers and Markets	124
Gathering Local Market Intelligence	129
Managing and Sustaining Team Performance	131
Enabling a Team Climate for Local Innovation and Collaboration	132
Bibliography	136

Part IV Multicultural Innovation: Leading Change from Planning to Execution

8 Facilitating and Orchestrating a Successful Transformation — 139
Moving from Global Reach to Global Readiness	140
Orchestrating Global Innovation Through Organizational Mechanisms	141
Optimizing and Facilitating Multicultural Team Collaboration	144
Connecting Global and Local Knowledge, from Concept to Market	149
Facilitating and Orchestrating Multicultural Innovation	156
Bibliography	159

9 The Future of Multicultural Innovation and Collaboration — 161
Practice: Optimal Solutions for Multicultural Collaboration	163
Research: Future Needs for Global and Multicultural Innovation	167
Bibliography	169

10 Conclusion: Ready for Your World Tour? — 171

Index — 175

List of Figures

Fig. 1.1	Challenges in conceiving and bringing new concepts to market	3
Fig. 1.2	The Multicultural Innovation Framework	8
Fig. 2.1	Global collaboration model	15
Fig. 2.2	The global leadership development journey	19
Fig. 2.3	Leadership behaviors for multicultural innovation	23
Fig. 3.1	Local participation in global innovation cycle phases	42
Fig. 3.2	Local team roles for the global innovation project	45
Fig. 3.3	Strategic co-creation through local to global integration	49
Fig. 4.1	Knowledge-sharing and cultural indicators	62
Fig. 5.1	Challenges in multicultural team collaboration	78
Fig. 6.1	Key values of a global innovation culture	101
Fig. 7.1	The global innovation cycle	115
Fig. 7.2	The global innovation project process	117
Fig. 7.3	Global launch road map and key milestones	118
Fig. 7.4	Multicultural collaboration project process	120
Fig. 7.5	Global team innovation climate	121
Fig. 8.1	Multicultural innovation framework in action	142
Fig. 8.2	Multicultural collaboration practices	148
Fig. 8.3	Connecting global and local knowledge	153

List of Tables

Table 4.1	Western and eastern cultural perspectives in knowledge-sharing for innovation	70
Table 5.1	Perceived challenges by global and local team leaders	81
Table 5.2	Trust and motivation behaviors for multicultural team innovation	88
Table 6.1	Organizational routines for a global innovation culture	105

List of Tools and Tips

Chapter 2	Global Teamwork Success Factors	20
Chapter 2	Global Leadership Development Audit	31
Chapter 3	Strategic Co-creation Audit	51
Chapter 4	Creating Global Dialogue Through Local Knowledge-Sharing	74
Chapter 5	Global and Local Team Development through Cross-cultural Learning	93
Chapter 6	Creating Collaborative Spaces Through Visual Communication	102
Chapter 6	Practices that Nurture and Sustain a Global Innovation Culture	110
Chapter 7	Knowledge-sharing Forums and Tools for a Team Innovation Climate	136
Chapter 8	Global Readiness Audit	158

List of Cases

Chapter 3	Nokia's Strategic Challenge: Moving from Global Implementer to Local Collaborator	42
Chapter 3	Philips Focus on Global Innovation and Design through Co-Creation	50
Chapter 4	Innovating Through Inclusive Dialogue and Culturally Diverse Teams at Intel	72
Chapter 5	Essilor's Cross-Cultural Lens for Viewing the World	89
Chapter 5	Connecting Adobe's Global Network and Empowering Team Collaboration	93
Chapter 6	Performance Excellence Through Cultural Diversity at Lenovo	104
Chapter 6	Siemens Convergence Creators: Global Knowledge Exchange and Collaboration as Key to Business Transformation	108
Chapter 7	From Global to Local Solutions: Wipro's Collaboration Focus	126
Chapter 7	Creating a Global Innovation Climate at Google, Twitter, and Airbnb: Insights from an International Product Leader	133
Chapter 8	Transforming How People Connect and Collaborate at Cisco	157

1

Introduction: Achieving Innovation in a Global and Dynamic Environment

A new concept is being prepared for a global launch, yet teams around the world are not onboard. The executive team in HQ has painstakingly developed the international strategy and proudly distributed the global plan to country managers worldwide. But the local and regional teams are protesting loudly or silently since they do not support the new strategy. After many delays for design and go to market requirements, the product is finally launched with great fanfare and fireworks around the world. It is expected to be a star performer that will accelerate international revenue. Yet customers are not connecting with the new concept and sales are far below target goals in local markets.

These are scenarios that I have observed all too often through my corporate, consulting and research projects. International organizations who are striving for mastery in global innovation have one challenge in common: effective collaboration across functions and cultures. If you're responsible for international programs, projects, products, or services, how are you facilitating innovation and collaboration around the world?

The inability of organizations to facilitate multicultural collaboration and knowledge-sharing can affect innovation in terms of concept design, strategic planning, marketing, operational efficiency, customer connection and sales performance in international markets. There are many stories of failed concepts, campaigns, and launches that have been re-told around the world, from hilarious translations to questionable product features…the list of concepts lost in translation grows every year. New innovations demand consideration of their cultural and local fit for consumers in both mature and emerging markets.

Yet multicultural collaboration is missing from the global innovation vocabulary. Current theory and practice tend to focus on culture and innovation as two separate categories that do not necessarily integrate in the international business environment. There are three common views: (1) culture is ignored by using universal innovation and leadership models that assume global application; (2) a dominant cultural perspective is applied due to research and practice views from one country or region; or (3) culture is treated as a separate topic in that we are told to understand and manage cultural differences rather than understand multicultural collaboration and optimize cultural diversity for innovation. Then where do we go from here?

The Evolving and Dynamic Global Business Environment

The only certainty in today's business world may be the changing demands for sustaining innovation in a dynamic, global marketplace. The growing role of emerging markets in international business growth has placed increased importance in understanding and responding to the particular needs of local consumers. While a global and centralized strategy may have been applied in the past, international organizations are now discovering that a local and decentralized approach to strategy-making can achieve increased responsiveness to international market opportunities. Emerging markets require more attention to cultural understanding and relationship-building in sharing and co-creating knowledge for front end innovation. Both mature and emerging markets represent culturally diverse consumers where organizations need to meet expectations for innovative solutions, time to market, and competitive products.

New technologies and the continuous evolution of connectivity, sharing, and data are driving the focus on knowledge and relationships. As international markets demand the design and delivery of localized products and services, multinational firms are facing increased pressure to optimize knowledge and innovate across the organization. The opportunity to access and share knowledge within the firm is determined by its international network. As Govindarajan and Gupta (2001) precisely observed, the benefit of an organization's global mindset derives from the ability to build cognitive bridges across local market needs and the company's own global experience and capabilities. The multicultural and networked business environment has created a growing need for knowledge-sharing between headquarters and subsidiaries. The ability to create products and services that respond to local market opportunities requires an effective knowledge-sharing process for geographically distributed teams.

There is pressure on companies to sustain innovation and rapidly respond to new market opportunities through the conception and execution of new solutions. In responding to global and local competition, organizations need to ensure a strong value proposition, increase international market share, and collaborate with customers, partners, and suppliers (see Fig. 1.1) around the world. This translates to a shorter and integrated value chain that can move efficiently from research to design to market. It requires organizations to optimize global and local team knowledge in key geographic locations in order to improve worldwide execution.

Can leaders and managers navigate the value chain and foster inclusive innovation for geographically distributed teams? New perspectives on global innovation show that organizations need to develop integrated innovation capabilities in optimizing a global footprint, collaboration, communication, and receptivity (Doz and Wilson 2012). With the impact of a digitally

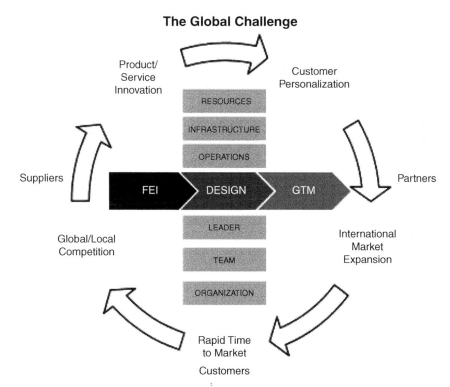

Fig. 1.1 Challenges in conceiving and bringing new concepts to market

connected network, there is also the consideration of live and virtual team collaboration using new technology platforms. In addressing organizational capabilities for effective execution, companies need to invest in relationship management capabilities in order to avoid cross-cultural challenges and issues that will block global network-centric innovation (Nambisan and Sawhney 2008). It is therefore important to consider the orchestration of organizational resources that strengthen collaboration and innovation management capabilities for conceiving and introducing new products.

As innovation is the driving force in the global economy, the different dynamics between mature and emerging markets need to be considered when introducing new concepts. Mature markets have long enjoyed front row seats during the globalization movement with accolades for high purchasing power, consumer demand, entrepreneurial success with competitive sectors, and highly developed infrastructures. The mature markets or developed economies include a majority of member countries in the European Union; North America/Canada and the US; Asia/Japan, South Korea, Hong Kong, and Singapore, as well as the Pacific including Australia and New Zealand.

In the meantime, emerging markets have become star players, achieving rapid economic growth and market success including countries in the regions of Asia, Latin America, Eastern Europe, Africa and the Middle East. Most notably, the BRICS countries (Brazil, Russia, India, China, and South Africa) have earned leading positions in the world economy followed by the next generation of high potentials that include countries such as Vietnam, Indonesia, Malaysia, Colombia, Turkey, and Nigeria. Although emerging market countries have recently experienced economic challenges, growth pains, and political turbulence, there is a dramatic shift in power structure and market growth in comparison to mature markets. Global growth was projected to decline to 3.1% in 2016, with growth in emerging markets and developing economies expected to strengthen slightly to 4.2% while growth in advanced economies declined to 1.6% (IMF 2016 WEO report).

On the other hand, developing economies are struggling to join the stage with the need for more access to education, technologies, and infrastructure to accelerate industrialization and growth opportunities in Africa, Asia, and Latin America. There is a focus on engagement of local communities, job creation, and capacity-building to address long-term needs for social and economic development in impoverished countries (US Aid 2015 and Asian Development Bank 2015). In addition to a slower average economic growth rate of 5.5% for least developed countries in 2014, the populations in the 49 poorest countries will double by 2050: Job creation is critical for preventing increased poverty, social unrest, and mass emigration (UNCTAD LDC Report, 2013, 2015). Local youth represent an untapped talent pool that is available to create solutions, learn new skills, and build businesses that make a positive impact on their communities.

As the global economy has evolved, new approaches to managing innovation have been introduced from open to reverse to social innovation. Open innovation assumes collaboration for generating internal and external ideas and concept creation within the wider eco system. Reverse innovation presents an opportunity to develop and test new ideas in emerging markets that eventually can be transferred to mature markets. Social innovation provides new solutions and value to problems faced in society and low income communities. For the purpose of this book, the focus is placed on global product innovation since there is a comprehensive cycle that moves from planning to execution, concept creation to global launch. The key phases demand organizational, team, customer and partner collaboration where the final concept is introduced to multiple markets around the world. Whether your teams are working with global, local, open, reverse, or social innovation, the projects will always demand multicultural collaboration.

As observed by many, we live in a volatile business environment where today's success story can become tomorrow's struggle or failure. Given the risks of international business, there will always be the need to respond and change in order to ensure organizational agility and performance. Commercial, financial, political, and cross-cultural risks always need to be considered. There is also an interesting difference in demographics where a predominantly young population represents emerging markets in Africa, Asia-Pacific, Latin America, and Middle East while there tends to be an older and more senior population found in mature markets within Europe, North America, and Japan. Attention should be paid to this demographic shift as it will influence the target markets and consumers of new products and services, in addition to the talent pool that is available to create, design, and deliver solutions around the world.

Accelerating Global Innovation Through Multicultural Collaboration

A global and rapidly changing business environment places an increased demand for multicultural innovation in order to ensure organizational performance and international market success. Leaders are facing the challenges of navigating a networked world where the need for collaboration and change are the only constants. International organizations have to strengthen collaboration capabilities for innovation projects and initiatives. This requires new competencies for leaders who will serve as orchestrators of global innovation across cultures and functions. Then how can leaders facilitate multicultural collaboration within the global innovation cycle? Consider your current readiness needs for the following phases that are critical to international and multicultural engagement:

Ideation – How is the creative process optimized for different cultural contexts and how are new ideas encouraged and shared between team members in local markets and headquarters?

Strategic Planning – How is your organization engaging in shared strategy-making through cross-cultural learning and knowledge-sharing?

Validation – How is your organization responding to cultural differences and local customer preferences when developing and testing new concepts?

Execution – How are you ensuring global team performance and success through recognition, transparency, and visibility for multicultural innovation and collaboration?

Culturally diverse views bring valuable insights to local problems and issues that are pertinent to solving international challenges. In today's constantly changing business environment, we need to pay more attention to local voices in order to orchestrate innovation around the world. Global innovation is thus transforming into a multicultural kaleidoscope with continuously changing patterns, shapes, and colors. How can you ensure a common vision for multicultural innovation within your launch plan? The answer lies within the leaders and teams that shape a multicultural and digitally connected network.

There are many books available on the topic of innovation, most address the principles and processes of innovation, while a few answer "what" is important for managing global innovation. In addition to building upon previous work, this book introduces a new perspective on the role of culture and "how" leaders can facilitate global innovation, from concept to market. By addressing both planning and execution needs, this book shows leaders how to facilitate and orchestrate innovation through multicultural collaboration.

This is an advanced book for leaders, managers, educators, and consultants who already understand the basic concepts of innovation, and are now seeking to understand how to develop successful innovation practices in an international, multicultural, and networked business environment. The book provides insights for leaders and teams who are responsible for global initiatives, consultants who advise organizations, and professors who have research and teaching interests related to international management, innovation, leadership, project management and team collaboration. It presents findings from qualitative empirical research that translate into new frameworks, models, and cases.

My research quest has been firmly focused on new leadership competencies and organizational capabilities that facilitate multicultural innovation and collaboration. The research was conducted between 2009 and 2015 through three international studies that involved semi-structured interviews conducted on-site and via phone, company visits and observations. Data collection involved a total of 200+ leaders from 25 different nationalities working for 45 multinational firms in 13 industries, with headquarters based in three world regions and 16 countries. In evaluating specific collaboration challenges and solutions, an extensive qualitative study and analysis was conducted from 2011 to 2014 with 105 global leaders in 36 MNCs with headquarters based in Asia, Europe, and North America (Global Innovation and Collaboration Study 2014).

The participants in the global study involving 105 leaders share a unique profile with participants in the first international study (2011). They are leaders responsible for orchestrating the global innovation cycle through the management and marketing of new products (from conceptualization to planning to execution and introduction to international markets). They are responsible for leading the global project with cross-cultural, cross-functional and geographically distributed teams. On site and telephone interviews were conducted anonymously with study participants who worked at the following multinational firms during the research period: Acer, Adobe, Alcatel-Lucent, Amazon, Apple, Applied Materials, BMW, Cisco, EMC, Essilor, Ericsson, Fiat, Ford, GE Healthcare, Google, Hitachi, HP, HTC, Hyundai, Infosys, Intel, Lenovo, LG, Mazda, Microsoft, Motorola, Nokia, Oracle, Philips, Renault, Salesforce, Samsung, SAP, Siemens, Symantec, and Toyota. The participants selected for both studies were primarily working in technology-driven industries (automotive, health, telecommunications and information communication technologies) since these sectors face increased pressures to compete globally, provide local solutions, and reduce time to market.

A regional study continued until 2015 including interviews with 40 local and regional managers based in the Asian subsidiaries of American and European multinational firms in the global study, specifically in China, India, Japan, Singapore, South Korea, and Vietnam. Since most of the leaders had cited challenges in collaboration between East and West, it was of great interest to compare and validate findings with regional and local leaders in Asia (Local Innovation and Collaboration Study 2015).

The book presents the findings and results of these studies along with the **Multicultural Innovation Framework** (see Fig. 1.2) which incorporates three collaboration drivers – Vision, Dialogue, and Space. It is organized into three sections where you will find key research discoveries, case examples, practice summaries, and audits for use within your current projects and initiatives. There are cases and examples from leading organizations including Philips, Siemens, Adobe, Airbnb, Google, Lenovo, Samsung and Wipro among others. The Multicultural Innovation Framework serves as a guide with visual models that provide examples for each section, offering insights to each collaboration driver. The intent is to provide readers with a guide that is informed by rigorous research yet highly applicable to organizational practices, allowing for an exploration and integration of the keys to leading multicultural innovation and collaboration.

With an overview of the global innovation landscape, **Chapter 1** provides an introduction to the book and a look at the changing forces for achieving innovation in a global and dynamic environment. **Part I** is focused on **Vision**

Fig. 1.2 The Multicultural Innovation Framework
Source: Dr. Karina R. Jensen, Global Minds Network, 2017

which involves the role of Global Leadership and Strategic Co-creation; Leaders are increasingly becoming knowledge facilitators and orchestrators while strategic co-creation becomes the primary process for identifying and creating new opportunities. **Chapter 2** presents the need for new leadership skills and behaviors in facilitating multicultural collaboration and orchestrating global innovation across cultures. **Chapter 3** presents the concept of strategic co-creation in order to optimize local knowledge and ensure successful strategies for international markets.

In **Part II, Dialogue** addresses the role of team communication where there is an increased need to engage through social networking, knowledge-sharing, and cross-cultural learning. **Chapter 4** identifies the primary cultural challenges in knowledge-sharing and presents diverse perspectives that impact front end innovation, from East to West. **Chapter 5** demonstrates the importance of dialogue and identifies solutions for strengthening trust-building and motivation during the project collaboration process, from ideation and planning to go-to-market and execution. The opportunity for leaders and teams to learn from local market knowledge can contribute to improved solutions for international markets.

In **Part III, Space** ensures an open and safe environment for collaboration through the essentials of a global innovation culture and team climate. **Chapter 6** demonstrates how a global innovation culture plays an important role in developing common values of cultural empathy, creativity, and collaboration. **Chapter 7** provides a practical view of the essentials of building an inclusive team climate within a culturally diverse and digitally connected business environment. The project collaboration process is central to team engagement since individual contribution is often linked to motivation and trust. The organizational culture values provide the foundation while the team climate enables the network and connection to international customers and markets.

In **Part IV, Multicultural Innovation** provides the guide to organizational change. **Chapter 8** presents a roadmap for optimizing multicultural innovation and collaboration. It demonstrates how the three levers of Vision, Dialogue, and Space influence organizational change, collaboration, and performance. **Chapter 9** looks to the future of organizational practices and research concerning multicultural innovation and collaboration, followed by **Chapter 10** with a conclusion and reflection on the global journey that awaits.

The purpose of this book is to show how leaders and managers can facilitate multicultural innovation and collaboration in order to strengthen organizational performance and international market success. It addresses the growing need for new leadership competencies and organizational capabilities for operating within a global business network. It presents a holistic framework that can assist leaders and their organizations to effectively conceive and execute global innovation strategies involving diverse teams, cultures, and markets. Through qualitative empirical research focused on multicultural innovation and collaboration, the book provides guidance for executives, directors, managers, educators and consultants who are interested in how to lead and facilitate innovation in international, dynamic and culturally diverse environments. It is with this intent that I hope you will benefit from new perspectives and models that respond to the changing needs of the global marketplace.

Bibliography

Asian Development Bank. *Development Effectiveness Review*, pp. 10–11, 2015.

Doz, Yves and Wilson, Keeley. *Managing Global Innovation: Frameworks for Integrating Capabilities Around the World*. Boston: Harvard Business Review Press, 2012.

Govindarajan, Vijay and Gupta, Anil. *The Quest for Global Dominance: Transforming Global Presence into Global Competitive Advantage*. San Francisco: Jossey-Bass/Wiley & Sons, 2001.

Jensen, Karina R. Accelerating Global Product Innovation through Cross-cultural Collaboration Study, Report 2011.

Jensen, Karina R. Global Innovation and Collaboration Study, 2014.

Jensen, Karina R. Local Innovation and Collaboration Study/Asia, 2015.

Nambisan, Satish and Sawhney, Mohanbir. *The Global Brain*. New Jersey: Wharton School Publishing, 2008.

United Nations Conference on Trade and Development. *The Least Developed Countries Report 2013*, "Growth with Employment for Inclusive and Sustainable Development." United Nations: New York and Geneva, p. 6, 2013.

United Nations Conference on Trade and Development. *The Least Developed Countries Report 2015*, "Transforming Rural Economies." New York and Geneva: United Nations, pp. 2–4, 2015.

US Aid Report. U.S. Agency for International Development, Research and Development Report FY 2015, Economic Growth, Education, and Environment, p. 21, 2015.

World Economic Outlook: Subdued Demand – Symptoms and Remedies, IMF, October, 2016.

Part I

Vision:
Global Leadership and Strategic Co-Creation

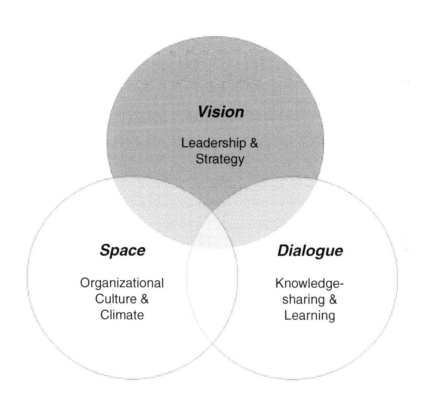

2

Leading Global Innovation and Collaboration

Responsible for a global portfolio across multiple business units, a senior global product manager at an international telecommunications firm had a full agenda managing multicultural teams across geographies and time zones. He had to develop and operationalize the global product strategy, roadmap, pricing and go-to-market/execution for an international product line. In order to ensure global project success, he emphasized the importance of developing a collaborative environment, in both live and virtual settings, where cross-cultural and cross-functional teams could exchange ideas. "It's about leadership through collaboration, capturing different perspectives. In almost every case, the top down approach or 'know what's best' never works," he paused, then added "what works is the more collaborative nature of getting different perspectives and integrating them in the creativity process." The senior manager believed that you need to optimize cultural diversity through a collaborative leadership style.

With a strong customer and relationship-focus, his group develops new customer solutions that respond to on demand business communication needs. This requires in depth knowledge of local markets and cultures, especially cross-cultural nuances and language differences. "Personal relationships and understanding of cross-cultural nuances are important. 'Yes' may not mean 'yes' in some countries in Asia. If a Frenchman says the customer is demanding, it may mean he's asking for something".

In order to improve cultural understanding when working with local teams and customers, a regular series of collaboration sessions encourage cross-functional and cross-cultural teams to share and create. There are

© The Author(s) 2017
K.R. Jensen, *Leading Global Innovation*,
DOI 10.1007/978-3-319-53505-0_2

regional forums that allow local teams to express their ideas, in addition to sharing updates and findings to upper management. "Global team leadership is about empowerment, making sure people feel empowered to bring change and influence the direction of the product," explained the senior manager.

Demands of a Multicultural Organization

The structures and systems of an organization largely determine its ability to recognize and adapt to international market opportunities. The complex challenge for today's organization is to create a holistic framework that connects knowledge between the organization and its customers worldwide. There is great hope for network-centric platforms that can rise above the organizational layers and open new communication channels. Yet, there's still the challenge of effective collaboration and knowledge-sharing when trying to communicate across geographic and cultural distances through the worldwide web. Engagement can be affected by different languages, communication styles, and cultural norms for sharing knowledge. When you're operating in multiple international markets, structures need to be simplified in order to allow for innovation and creativity across countries.

In order for leaders to effectively orchestrate global innovation, they will need to consider organizational capabilities and resources. As shown in the Global Collaboration Model, there are five organizational mechanisms that influence multicultural collaboration – leadership behaviors, innovation strategy, knowledge-sharing structure, communication vehicles, organizational culture and team climate (Jensen 2015). These mechanisms can influence team motivation and engagement during project collaboration through innovation routines (ideation, planning, validation, and execution) that determine global performance. They can either accelerate or block performance, as measured by time to market, concept localization, customer demand, and local sales results.

In order to better understand the framework and role of organizational capabilities, let's take a closer look at the mechanisms that can influence collaboration and performance (Fig. 2.1). The *innovation strategy* plays a key role as the focus on global or local co-creation and planning can determine team engagement. The *knowledge-sharing structure* is determined by team roles in planning and executing the innovation strategy. *Communication vehicles* can shape collaboration through the particular balance of face-to-face meetings and virtual communication platforms via

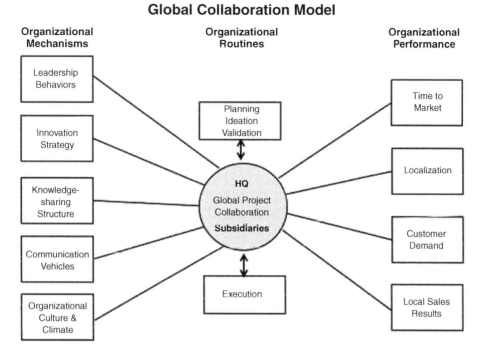

Fig. 2.1 Global collaboration model
Source: Dr. Karina R. Jensen, 2017

mobile, web, and video. The frequency and intensity of organizational communication flow then influences the local and regional roles of the multicultural team members. Finally, an environment that can optimize cultural diversity requires an *organizational culture* and *team climate* that can nurture and sustain multicultural innovation and creativity.

These organizational mechanisms have been examined and validated for their particular influence on behaviors of knowledge-sharing and collaboration involving leaders and cross-cultural teams. Leaders who can orchestrate these mechanisms as an interdependent process can greatly influence team motivation, project success, and organizational performance. Since the leadership role is critical to influencing collaboration, this book will explore the organizational mechanisms and leadership behaviors that effectively facilitate multicultural collaboration during the global innovation project process.

Global team performance and market success require a balance between internal and external drivers. In meeting performance measures, there are organizational pressures in timely market delivery, achieving cost and revenue objectives, and ensuring product and service quality. In addition, there are growing market pressures for meeting internationalization and localization needs, ensuring customer satisfaction, keeping up with local competition, and meeting international sales goals (Jensen 2014). These performance expectations highlight the importance of organizational responsiveness to global and local market demands. This requires a leader that is capable of accelerating organizational performance and international market success across functions and cultures.

The Global Leadership Development Journey

In order to understand global leadership, it would be wise to re-visit the general definition of leadership which is focused on creating a vision and providing guidance. A definition that speaks to the roles of indirect and direct leadership through the power of communication is "an individual who significantly affects the thoughts, feelings, and/or behaviors of a significant number of individuals" (Gardner 2011). Leadership involves the exercise of influence and is linked to outcome. Management, on the other hand, places a focus on setting and achieving goals through execution. With today's complex demands and organizational needs, there is also the demand for managerial leadership which requires elements of leadership and management – vision, strategy, and execution – that incorporate interpersonal, influencing, and goal-setting skills.

Leadership has been the source of great fascination and research over the decades with numerous theories that have evolved. The key theories that have shaped leadership studies include behavioral, contingency, transactional, and transformational. Behavioral theories revealed the ability to learn and develop leadership skills involving tasks and people. Contingency theories then examined how leadership styles depend on certain situations, Finally, transactional and transformational leadership theories compared individual and team motivation; transactional tends to focus on individual achievement through structure and operational efficiency while transformational is focused on team achievement through new ideas and common interests.

Emerging theories and practices are increasingly focusing on values and competencies for inspiring trust, motivation, and collaboration in leading within organizations. A model that has provided a valuable foundation for examining these leadership skills is the introduction of Emotional Intelligence (EI) as a stronger indicator of leadership performance in comparison with IQ or technical skills (Goleman 1998, 2004). The five skills that maximize leadership performance include self-awareness, self-regulation, motivation, empathy, and social skills that can be strengthened through persistence, practice, and feedback (Ibid). The EI skills contribute to the development of skills that support your core values and the foundation for leadership performance.

The primary differentiation within leadership models relevant to team collaboration is the role of power; the degree that leaders hold onto power and decision-making authority compared to their ability to empower teams through shared decision-making. It is the difference between autocratic and democratic leadership styles. Supporting transformational theories and team empowerment, the Kouzes and Pozner (2002) model demonstrates five specific characteristics for successful leadership that include being a role model, inspiration, facing adversity, empowerment, and generating enthusiasm. It is a model that supports elements of leading change and innovation within the organization. Depending on the leadership characteristics and practices that appeal to your core values, it is important to explore and identify models that speak to your ethos.

In order to fully prepare for multicultural team collaboration, there is an emphasis on developing cross-cultural competencies through global leadership models. Intercultural research and theory has helped link general leadership theories to the challenges of leading in a global context (Gundling et al. 2011). The most common approaches are focused on cultural dimension models, in order to recognize and understand cultural differences through country comparisons. There is the classic Hall's Low Context and High Context model (1976), the pioneering frameworks of Hofstede's Six Dimensions model (1997, 1980, 1991) and Trompenaars' Seven Dimensions model (1999), the Globe Study (2004), and most recently Meyer's Culture Map (2014). Cultural dimension models provide a good foundation for developing knowledge about cultural differences and local practices when conducting business around the world. However, multicultural business contexts demand an additional layer of competencies and behaviors in order to effectively interact with multiple cultures in rapidly changing contexts.

Leaders and managers increasingly need to relate to a global and networked environment that requires interactions with colleagues from a variety of backgrounds and cultures. Cultural conflicts can arise from project collaboration where it's important to consider a cultural sense-making approach for evaluating and responding to situations, in addition to value dimensions and communication styles (Bird and Osland 2005). Cultural Interaction models apply specific cognitive skills and behaviors that help interpret and navigate cultural situations. Placing more focus on cross-cultural interactions, there is the Participative Competence model that promotes the facilitation of interactive translation and knowledge-sharing activities (Holden 2002). The Cultural Intelligence model applies knowledge, mindfulness, and behavioral development in cross-cultural situations (Inkson and Thomas 2004). The Cross-cultural Quotient involves your ability to interact effectively across cultures and reflects your capability to gather, interpret, and act within multicultural settings (Earley and Ang 2003). The Global Mindset model addresses leadership competencies in openness, knowledge, and integration of diverse cultures and markets as measured in intellectual, psychological, and social capital (Govindarajan and Gupta 2001, Javidan et al. 2010). All of these models address cognitive skills, behaviors or competencies applied to cross-cultural interactions.

While there's a growing interest in theoretical and empirical research on collaboration, the role of cross-cultural collaboration within the context of global innovation has received limited attention. Collaboration from a cross-cultural perspective can be defined as the study of similarities and differences in the processes and behavior at work across different cultures (Gelfand et al. 2007). Academic and business literature still demonstrate a strong emphasis on studying cultural dimensions in order to recognize and manage cultural differences rather than understanding differences and similarities in order to optimize multicultural collaboration. There are some emerging perspectives such as the intracultural competence model which addresses the capability to leverage cultural and/or ethnic diversity within teams for business advantage (Trompenaars and Hampden-Turner 2012). There is also the importance of leadership learning through team collaboration. Through the process of leading and managing multicultural teams, leaders can refine and develop the knowledge and skills required for global leadership (Maznevski and DiStefano 2000). It is an interdependent cycle that allows leaders and teams to share knowledge and apply multicultural learning to project collaboration.

With increased demand for global collaboration, intercultural interactions through cultural synergy (Boyacigiller 2002, Adler 1983) are highly relevant to theory and practice in multicultural innovation and collaboration. Cultural synergy shows under which conditions universal (patterns common to all cultures) and pluralistic (culturally specific patterns) approaches can be used. The Multicultural Innovation Framework introduced in this book thus applies a cultural synergy approach that allows you to identify universal practices as well as particular cultural considerations for leading within the context of multicultural innovation and collaboration.

What is most important is to explore your global leadership development journey according to four layers: core leadership values, cultural dimensions,

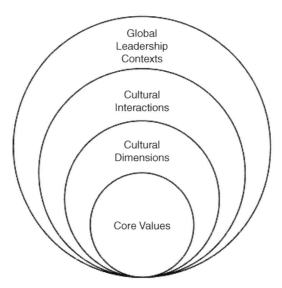

Fig. 2.2 The global leadership development journey

cultural interactions, and global leadership contexts (see Fig. 2.2). Each development stage is essential to developing and applying new knowledge and cognitive behaviors. There are leadership models and practices available for your specific professional development needs at each stage. Core leadership values lead to greater self-knowledge that guide the development of skills and competencies; cultural dimensions provide insights to cultural differences in team and organizational settings, and cultural interactions allow you to identify and develop cognitive behaviors for facilitating interactions with cross-cultural teams.

These three layers are then applied to leadership and business contexts in order to understand how to effectively lead and manage international projects and initiatives. Understanding cultural dimensions and cultural interactions strengthen cross-cultural competencies, however, there is also the need to manage within specific business contexts. Knowing cultural profiles and practicing cultural interactions are only as effective as their integration with leadership practices for inspiring concept creation, designing business strategies, managing the project and team process as well as facilitating communication across regions. This book addresses leadership within the particular context of leading multicultural innovation and collaboration. It requires facilitation of knowledge-sharing with cross-cultural teams and the orchestration of the global innovation cycle and project process. It's a continuous and holistic learning journey that empowers and enables leadership for multicultural innovation.

Optimizing Multicultural Team Collaboration

The global project leader serves an important role in communicating the vision and providing the road map for cross-cultural and geographically distributed teams. Global team leadership requires openness to diverse cultures and perspectives. There is a greater need for trust-building and frequent communication throughout key project phases, from ideation to execution. As noted by a global programs director, "It's important to listen to people, to get the feeling of how it's going for everyone, important to let people express themselves everywhere, not to make it a privilege or priority for particular members or countries." The project leader needs to ensure the time and space for international teams to share knowledge and exchange ideas.

The opportunity to learn about local practices and cultural perspectives requires face-to-face interactions with an emphasis on relationship-building. In order to engage in front-end innovation around the world, an open and safe environment with a collaboration focus is important for cross-cultural and cross-functional teams. Team members need to build trust in order to share, discuss, and resolve problems. Cultural understanding and team cohesion are improved by offering learning and knowledge-sharing opportunities in a consistent manner.

The management of a geographically distributed team is not an easy task as it requires the full engagement and commitment of team members that are dispersed across countries and time zones. This places particular importance on the role of the global team leader as knowledge facilitator and project orchestrator. It is a multidisciplinary role that demands effective ideation, strategic planning, project management, and team leadership across functions and cultures. As shown in below summary, global team success requires communication and cross-cultural understanding, team engagement, efficient project collaboration and the development of a shared vision that will propel the team from concept to market in every geographic location (Jensen 2011). These factors will be addressed in more detail in the remaining chapters.

> **Global Teamwork Success Factors**
> 1. Understand cultural and market differences.
> 2. Practice frequent and open communication on global and local levels.
> 3. Engage global teams through effective communication and collaboration technologies and tools.
> 4. Establish clear goals and objectives through effective project planning.
> 5. Develop a shared vision that ensures a common understanding.

> 6. Focus on relationships through socialization and personal interaction.
> 7. Ensure openness and flexibility in adjusting to cultural differences.
> 8. Establish trust across cultures and regions.
> 9. Develop cultural empathy through respect and understanding.
> 10. Create a common language.

In navigating the global innovation cycle, the greatest challenges facing leaders and senior managers in facilitating knowledge-sharing and collaboration include four main areas – **global market visibility**, **strategic planning process**, **local market intelligence**, **team communication and knowledge-sharing** (Jensen 2014). *Global market visibility* shows the lack of access to information and dialogue between local subsidiary offices and headquarters due to competing priorities. The *strategic planning process* points to the difficulties of capturing local needs and providing sufficient feedback for effective planning and execution. *Local market intelligence* is complicated by the lack of knowledge and skills necessary to understand local customers, markets, and cultures. *Team communication and knowledge-sharing* refer to the challenges of managing on site and online communications that involve cultural and language differences.

Leaders are challenged by the ability to motivate and collaborate with team members across cultures and geographies. Trust-building is viewed as the top challenge in team collaboration where project management pressures and lack of effective team processes and communication tools can reduce trust amongst geographically distributed team members (Jensen 2011). There are few opportunities to pursue dedicated creativity and brainstorming sessions since there is often a lack of time and space. Due to less verbal interaction and more online communication, there is a greater need for relationship-building, clearer objectives and procedures, increased communication, as well as practices and tools available for group collaboration.

New Leadership Model for Multicultural Innovation and Collaboration

Global innovation and collaboration demand particular leadership competencies in a multicultural and networked environment. Leaders need substantial cultural and market intelligence, facilitation, and orchestration skills in order to accelerate innovation and performance around the world. Specific leadership competencies are required to inspire a common global vision, engage multicultural and multifunctional teams in dialogue, and orchestrate a dynamic space for effective project collaboration and execution. Yet leadership competencies have not been examined for this highly challenging and demanding environment where performance is critical to international market success. There is limited insight on applying specific behaviors to the particular situational context when faced with leading global innovation projects and initiatives.

Leadership behaviors can have great influence on the outcome of the global innovation project. For example, a senior manager working for a leading US-based multinational computer company had invested much time with the executive team in building strong relations with teams in key international markets. However, the situation completely changed when a new executive replaced the outgoing leader. "We experienced a recent re-organization where the executive team members switched hands," he explained: "In the past, the previous vice president was very pro-active in getting local buy-in and feedback." The new executive's approach focused on a top down leadership style where he wanted HQ to lead and control the effort which resulted in a lack of open feedback and limited engagement from local teams. As a result, the senior global launch manager faced new challenges in trying to motivate the international teams, explaining that "if you want to be creative as a global company down the road, it's important to encourage ideation from your international counterparts instead of focusing on the HQ team."

In order to effectively lead global innovation, new behaviors for global team leadership have been explored and identified through the Global Innovation and Collaboration Study (Jensen 2014). Specific team leadership behaviors are necessary for effectively facilitating multicultural team collaboration during the global innovation cycle, from concept to market. There are certain behaviors that are more suitable and effective for each of the stages in leading creation, planning, validation, and execution. The four leadership behaviors are focused on the key values of **empowerment, inclusion, direction, and communication** (Jensen 2014).

In developing a common vision and motivating the team to initiate and create new ideas, *empowerment* creates unity, engagement, and creative inspiration. *Inclusion* invites engagement and contribution through a cross-cultural mindset, collaboration, and relationship-building. *Direction* provides authority and guidance to meet integration and validation needs for making informed decisions. *Communication* is focused on interpersonal interactions and relationship-building through consistent dialogue and active listening.

Fig. 2.3 Leadership behaviors for multicultural innovation

The leadership behaviors integrate with the global innovation cycle and they adapt to the changing needs of project phases and cultural contexts (see Fig. 2.3). The ***empowering leadership behavior*** provides inspiration and support for the creation phase through the development of a common vision worldwide, engaging everyone in the development of new ideas and solutions. In achieving agreement upon project objectives, influencing and engagement skills are essential for motivating team members in order to drive results while ensuring sufficient support and resources. The ***inclusive leadership behavior*** invites collaboration through relationship and trust-building. It is the ability to demonstrate cultural empathy by understanding and optimizing diverse perspectives. This facilitates the ability to capture global and local market intelligence through knowledge-sharing and cross-cultural learning practices during the strategic planning phase.

In order to effectively evaluate and integrate knowledge for concept validation, the ***directive leadership behavior*** applies a decisive process with an execution focus that is valuable to the validation phase. In order to orchestrate effective execution and sustain team performance, the ***communicative leadership behavior*** nurtures open dialogue through active listening and knowledge-sharing, connecting team members in the global network throughout the project collaboration process. While these four behaviors are critical to multicultural collaboration throughout the global innovation cycle, they are especially suitable for the leadership competencies needed in particular phases, from creation to execution.

Creation Phase and Empowering Leadership Behavior

The leadership behavior identified for effectively facilitating multicultural team collaboration during the creation phase is empowering leadership where the project leader creates a common vision, engages the team, and inspires new ideas and solutions. In achieving agreement on project objectives, influencing and engagement skills are essential for motivating team members to reach the desired results. It requires the development of a collaborative environment where team members feel safe and open to sharing knowledge through initiative and communication. "You need to create the circumstances and ensure that people are being rewarded and moving towards the same target," explained a director, while a senior manager from another firm emphasized that "collaboration happens during the concept creation phase where we typically need a variety of ideas, voices from various perspectives in all geos." There is the necessity to develop the vision and an environment where team members are free to create and exchange knowledge and ideas that support the objectives for the global innovation project.

In providing space and opportunity, the empowering leadership style encourages team members to initiate and take ownership of the project process. Engagement of team members is especially important in the creation phase where local market knowledge and new ideas can contribute to project success. As noted by a senior manager: "The early phase needs the right brain and more creative approach to influence and inspire where you need to engage and influence the team in addition to getting support and approval from top management." When collaborating with team members of diverse cultures, it becomes critical to recognize their contribution and level of involvement in front end innovation and throughout the global innovation project cycle.

"Global team leadership is empowerment," noted a senior manager responsible for global product management and marketing, "It's about empowering and delegating, making sure there are regional forums that allow people to express ideas and bring forth to the table, as well as opportunities to share updates and findings to upper management." Leaders need to consider how to design an open space for initiating creativity and idea generation followed by a series of regular team sessions devoted to collaboration and knowledge-sharing throughout the global innovation project.

Given the differences in communication styles and English language ability, additional time, respect, and patience are required for achieving understanding and collaboration. As emphasized by a director, "the team leader should be the one who gains visibility for the team from upper levels, who needs to ensure that important work is visible and that members feel recognition, to see the innovation moving through the pipeline and understand their value and contribution to innovation, to understand their part in the big picture." Empowering leadership requires the project leader to create inspiration and motivation by recognizing the value of knowledge and talent provided by cross-cultural team members and rewarding them with project collaboration roles from the front end of innovation to execution and go-to-market.

Planning Phase and Inclusive Leadership Behavior

When moving into strategic planning and co-creation, the inclusive leadership behavior is valuable for developing cultural understanding and relationship-building within cross-cultural teams. Openness to new cultures and an appreciation for different perspectives shape the mindset for cultivating inclusive leadership. In order to connect with local customers and markets, it's important to develop cultural empathy and optimize knowledge-sharing and learning with local teams. As expressed by a global product planning director when reflecting upon his own experience: "Be open-minded, be open to others world view. Don't be too focused on your own objective and solution and don't push your own agenda too much. If you need to achieve certain results, it works better to see through other's eyes." Moving from global planning to strategic co-creation allows improved focus on international markets and customer needs rather than only focusing on a global view from headquarters.

Cultural differences and diverse views may sometimes lead to conflict where the team leader needs to understand how to facilitate and mediate issues that occur between team members. There is a need to set expectations in terms of the global strategy, new concept development, and market priorities. and ensure understanding within the team in order to ensure contribution of critical knowledge. The strategic planning process requires a deeper view of local dynamics and practices where dialogue requires an effective feedback process. As expressed by a senior manager: "It's very productive to have collaborative team management where you try to capture everyone's feedback since it's important to understand hidden gems where you need to dig deeper."

Inclusive leadership behaviors support a collaborative approach that invites participation from team members. The focus on inclusion and contribution to the global launch project demands respect and openness to other cultures. "You need to show respect and imagine their perspective and world view. Show that you understand them and you can relate to them," advised a director of international products in reference to working with local team members. Contextual sensitivity is especially important when working across geographic and cultural distances in order to ensure cultural empathy. There is an opportunity to capture unique insights when there is openness and sensitivity to the value of cross-cultural team knowledge for the global innovation project.

Trust requires relationship-building and cross-cultural learning in order to increase communication and interaction. When developing and co-creating innovation strategies, there are differences in knowledge-sharing, communication, and strategic views where a project leader needs to be flexible and adapt to diverse values and perspectives. "What makes leadership complicated is not only time zone differences, cultural differences, and how you deal with teams that don't perform…you need to ensure teams grow together…you need to develop trust-based relationships," emphasized a senior manager. Another senior product manager also noted that a collaborative style gives team members the opportunity to show and demonstrate their own leadership: "Since the launch is very structured, one needs to give more space for members to initiate and provide leadership for new ideas." By focusing on openness and respect for culturally diverse team members, there is an opportunity to build trust and stronger relations during the strategic co-creation process. Trust, respect, collaboration, and cultural empathy become the key elements of inclusive leadership.

Since relationship-building and face-to-face interactions are important to building trust and collaboration, live or on-site planning sessions are the best vehicles for bringing global teams together in one location. Face-to-face meetings combined with various communication technologies can sustain dialogue post-planning sessions. A senior product manager noted that "Face-to-face meetings are a key element, especially when you're new in this position and starting a project; it's important to meet and show, it's about you going to them and its key that members understand how they're part of a global approach and that they feel they're part of a global team." It's important that everyone has the opportunity to bring their ideas and their voices to the strategic planning phase whether in person or online.

Validation Phase and Directive Leadership Behavior

The directive leadership behavior is focused on providing authority and direction for the team during the global innovation project. It's valuable during the validation phase since it requires direction with integrative decision-making. This behavior is associated with strong leadership from concept to execution combined with a firm and structured project management process. Directive leadership should ensure a balance of vision and guidance rather than serve as an authoritarian style with top-down directions. If it becomes strictly a top-down approach where strategy and process are determined by the senior management team, it will be difficult to engage team members during the execution phase. There will not be much involvement in the project phases since the project is driven by a central decision and plan where the focus for local team members is to provide support and collaboration for successful execution.

A directive leadership style that allows for more involvement during project phases yet retains leadership in making key project decisions is more effective for multicultural team collaboration. The validation phase requires more directive leadership to screen ideas, evaluate concept feasibility, and determine the most effective solution for go-to-market and the global launch. The business case needs to demonstrate market potential, customer demand and a clear value proposition in key geographies. As the project moves into the execution phase there will be a need to balance directive and communicative leadership styles in order to ensure effective project management and a successful outcome. As noted by a senior manager responsible for the worldwide product business at his company: "You need to be decisive, to have the ability to review different kinds of information and knowledge, manage conflicts and needs, and then determine the best solution." While the project leader needs to ensure collaboration and consensus, it is also necessary to make a decision that represents the needs and priorities of the global innovation project. Thus directive leadership should emphasize decisiveness and direction when there is the need to drive the process forward for specific phases of the global innovation cycle.

If there's a focus on leadership directives rather than team participation, there may be less interest and motivation to contribute to project collaboration. A senior manager emphasized, "You can't be autocratic, you will lose people. Don't tell them what you're doing only…include people, and involve them based on their contributions." In addition to a strong directive focus, there is also the challenge of using the cultural self-reference criterion for the manager leading the innovation project where one's own set of cultural values and perspectives are applied. Leaders need to be careful in not relying on their own cultural approach or the home country of MNC headquarters. The directive leadership style should demonstrate a delicate balance through engagement and integrative decision-making while ensuring sufficient direction and guidance for the planning, validation, and execution phases.

Execution Phase and Communicative Leadership Behavior

The communicative leadership behavior applies interpersonal communication skills through formal and informal interactions on global and local levels. It is the most important leadership behavior in facilitating knowledge-sharing throughout the global innovation cycle, with a particular emphasis on the execution phase and effective project collaboration. The communicative leader emphasizes active listening skills and attention to language use. It is essential to listen to team members and allow sufficient exchange and communication during the global innovation project. Taking time to listen to diverse perspectives and local points of view can help the project leader improve understanding of particular needs or requirements for local markets. As noted by a senior manager, "you need to be a good listener; you're not a local expert so you need to listen to the locals."

There is also the need to have regular communication that involves face-to-face meetings as well as electronic communications in order to sustain a dialogue between the project leader and the team. In order to ensure that everyone is informed and engaged, there are several modes of communication available such as web sites, portals, or social media network sites. This demands the ability to foster productive conversations where every team member is involved and shares their perspective. The setting of expectations and sharing of context need to be communicated clearly and precisely. Due to the interaction of team members from diverse cultures, there is a greater need to listen and ensure understanding through an awareness of local languages and the preferred modes of communication.

The global project leader serves as the central contact for the innovation project and team communication. This places more importance on the leader's influencing skills for increasing interaction and collaboration amongst cross-cultural team members. Leaders serve as knowledge facilitators and connectors to resources and people in driving the internal network. When addressing multicultural collaboration, a senior manager explained that "it's being connected between all of them, behaving like a network, where the manager is seen as a node in the network who's heavily involved in all communications and can connect people." In order to ensure successful execution, there is a focus on continuous communication and exchange for enhancing knowledge-sharing in a global and digitally connected environment.

Leadership Behaviors in Asia

Leadership behaviors need to consider the cultural and business context. When evaluating the preferences of local management teams in Asia, the four behaviors of inclusive leadership, communicative leadership, directive leadership and empowering leadership were supported (Jensen 2014). However, the behaviors related to directive and inclusive leadership received special emphasis since authority and direction were valued as much as relationship-building and collaboration for front-end innovation. Moreover, communicative behaviors using interpersonal communication skills through active listening with attention to cultural practices are important to building credibility with local teams. Finally, the empowering leadership style appears to receive more interest with local team members as local and global markets become more competitive and require new ideas and customer solutions.

Local study participants in Asia emphasized the growing need for encouraging initiative and exploration of new opportunities. Most of the Asian cultures studied, including China, Japan, and South Korea, do not tend to take initiative unless there is a directive or a request involved. Although initiative and creativity are not common business practices, they are receiving more attention due to the innovation needs of the global marketplace. As expressed by a senior manager based in China: "We need a different mindset, the company is managed with too much of a systematic approach…We need to allow risk and failure. Everybody avoids risk, they want to play it safe. We need to allow some part of organizational resources devoted to this area." Inspiring new ideas and solutions during the creation phase motivates and encourages local team members to innovate and take more initiative. A regional executive based in Singapore emphasized that "early on you need to be very inclusive and towards the end focus on ownership and execution." Local team members need the opportunity and support to take initiative and create new ideas.

In reviewing the kind of team leadership behavior necessary for effectively leading global innovation and facilitating multicultural collaboration, leaders should consider behaviors that encourage empowerment during the creation phase, inclusion for the strategic planning phase, direction for the validation phase, and communication for the execution phase. When applied to the context of the global innovation cycle, the four behaviors strengthen the capabilities of global leaders to facilitate multicultural collaboration.

During the creation phase, empowering leadership creates a common vision that transcends cultures and functions while inspiring the creation of new ideas and solutions. Inclusive leadership allows an open and collaborative approach to working with team members from diverse cultures during the strategic planning phase. Directive leadership ensures guidance and decision-making for the validation phase involving the new concept and business case, including the transition to project execution. In order to ensure effective project collaboration, communicative leadership sustains the dialogue necessary for successful execution through active listening and attention to cultural differences in communication and knowledge-sharing.

Global Leaders as Facilitators and Orchestrators

When I participated in an international conference and consulting assignment in Jakarta, a local professor and colleague turned to me and inquired whether I knew the meaning of leadership in Indonesia. Since I was eager to learn, I asked her to please enlighten me so that I could understand local beliefs and values. "In Indonesia", she explained, "there is the concept of informal leadership where the leader does not have a formal title or responsibility, however, he or she has earned the respect of the community. The leader serves the role of connector who has influence and authority in decision-making." This discovery reminded me of the emerging roles of global leaders as connectors and knowledge facilitators.

In order to sustain a collaborative dialogue between cross-cultural teams, the global project leader serves an important role as knowledge facilitator while orchestrating the global innovation project process. In facilitating multicultural collaboration, leaders have the opportunity to listen, recognize, and respond to knowledge shared by team members. Sharing knowledge between cultures requires special attention to the diverse perspectives, ideas, and values of team members worldwide. Leaders and teams also need to consider communication vehicles that emphasize face-to-face interaction to facilitate trust and relationship-building where communication technologies sustain continued interactions throughout the global innovation process.

In facilitating knowledge-sharing and learning in the front-end, leaders also need to address orchestration of the global project collaboration process. Paying attention to innovation leadership behaviors within the project phases can accelerate multicultural team performance. This requires empowering leadership behaviors for inspiring and generating new ideas in the creation phase to inclusive leadership for connecting local and global knowledge in the strategic planning phase. Then there's the validation of the concept and business case through directive leadership followed by project collaboration and execution through communicative leadership behaviors. From the global value chain to the global innovation cycle, leaders need to facilitate and orchestrate an integrated process across functions and cultures.

> **Global Leadership Development Audit**
>
> **Core Values**
> - What are your current leadership strengths?
> - Which leadership models and practices do you support?
> - How would you summarize your core values in leadership?
>
> **Cross-Cultural Competencies**
> - What are the cross-cultural competencies required for your current projects?
> - What are the key challenges for team communication and collaboration?
> - Do you have an understanding of the cultural dimensions within your team?
> - Have you reviewed and applied a cultural interaction model for your current project?
>
> **Multicultural Innovation Context**
> - How are you applying leadership behaviors throughout the global innovation cycle?
> - In which global innovation phase are you experiencing multicultural collaboration challenges?
> - How could you improve your leadership for specific phases in the global innovation cycle?

Bibliography

Adler, Nancy J. "A Typology of Management Studies Involving Culture", *Journal of International Business Studies*, Fall (1983): 29–46.

Bird, Allan and Osland, Joyce S. "Making Sense of Intercultural Collaboration", *International Studies of Management and Organization*, 354: (Winter 2005–2006): 115–32.

Boyacigiller, Nakiye A. et al. "Conceptualizing Culture: Elucidating the Streams of Research in International Cross-cultural Management". In B.J. Punnett and O. Shenkar (Eds.) *Handbook for International Management Research*. Ann Arbor: University of Michigan Press, 2002.

Earley, Christopher P. and Ang, Soon. *Cultural Intelligence: An Analysis of Individual Interactions Across Cultures*. Palo Alto, CA, USA: Stanford University Press, 2003.

Gardner, Howard E. *Leading Minds: An Anatomy of Leadership*. New York: Basic Books, 2011.

Gelfand, M.J. et al. "Cross-cultural Organizational Behavior", *Annual Review of Psychology*, 8(2007): 479–514.

Goleman, Daniel. "What Makes a Leader?" Best of HBR 1998, *Harvard Business Review*, January 2004.

Govindarajan, Vijay and Gupta, Anil. *The Quest for Global Dominance: Transforming Global Presence into Global Competitive Advantage*. San Francisco: Jossey-Bass/Wiley & Sons, 2001.

Gundling, Ernest et al. *What Is Global Leadership?* Boston/London: Nicholas Brealey Publishing, 2011.

Hall, Edward T. *Beyond Culture*. Garden City, NY: Anchor Press, 1976.

Hofstede, Geert. *Culture's Consequences: International Differences in Work-Related Values*. Beverly Hills, California: Sage Publications, 1980, 1991.

Hofstede, Geert. *Software of the Mind: Intercultural Cooperation and Its Importance for Survival*. McGraw-Hill: New York, 1997.

Holden, Nigel J. *Cross-cultural Management: A Knowledge Management Perspective*. Great Britain: Prentice Hall, 2002.

House, Robert J., Hanges P.J., Javidan. M., Dorfman, P. and Gupta V. *Culture, Leadership, and Organizations: The GLOBE Study of 62 Societies*. Thousand Oaks, CA: Sage Publications, 2004.

Inkson, Kerr and Thomas, David C. *Cultural Intelligence*. San Francisco: Berrett-Koehler, 2004.

Javidan, Mansour, Teagarden, Mary and Bowen, Dave. "Making It Overseas", Harvard *Business Review*, (April 2010): 109–13.

Jensen, Karina R. Accelerating Global Product Innovation through Cross-cultural Collaboration Study, Report 2011.

Jensen, Karina R. Global Innovation and Collaboration Study, 2014.

Jensen, Karina R. "Global Innovation and Cross-cultural Collaboration: The Influence of Organizational Mechanisms," *Management International*, 19 (2015): 101–16.

Kouzes, James M. and Posner, Barry Z. *The Leadership Challenge*. San Francisco: JosseyBass, 2002, 2012.

Maznevski, Martha L. and DiStefano, Joseph J. "Global Leaders Are Team Players: Developing Global Leaders through Membership on Global Teams". *Human Resource Management*. 39: 2/3(2000): 195–208.

Meyer, Erin. *The Culture Map*. New York, NY: Perseus Books, 2014.

Trompenaars, Fons and Hampden-Turner, Charles. *Riding the Waves of Culture: Understanding Diversity in Global Business*, New York: McGraw-Hill, 1999, 2012.

3

From Global Strategy to Strategic Co-creation

Based in the US subsidiary of a technology firm with headquarters in Asia, the global program director looked forward to working on a new product introduction. The US business unit would have global responsibilities in order to accelerate the decision-making process for going to market. However, the product group was still located in Asia HQ where all design and development decisions were made, without the collaboration and involvement of the regional management. During the concept creation phase, the global program director and his team eagerly shared ideas and plans for North America and international regions, including feature requests. The executive team in Asia HQ received the ideas without providing feedback as they developed and confirmed a global plan that was best suited for capabilities in HQ.

"People like to be noticed, to bolster value to the region and the company. If they think global HQ will steal their idea or use it as their own idea, they will be hesitant to share," noted the global program director as his teams lost enthusiasm due to their lack of involvement in strategic planning. A few months later the executive team in the Asia HQ announced that all product and marketing activities would be centralized in order to manage the complete innovation cycle. "This changed the company culture to one of control where you couldn't confront or bring new questions to the table," explained the program director in a frustrated voice.

The strong hierarchical control and the inability to openly participate in ideation and planning was not appreciated by the international management team based in the US, since they were accustomed to a collaborative project process. Soon thereafter several members of the international team left the company in search of new opportunities. Unfortunately, this resulted in a delayed product development schedule and the launch of a product that had mixed performance results in the North American and international markets.

Innovation Strategies for Mature and Emerging Markets

The global marketplace that drives innovation today is experiencing major shifts through the strategic dimensions of context, content, and process. The context for strategy-making is heavily influenced by globalization, customization, and collaboration where knowledge and innovation move easily and quickly across borders (Davenport et al. 2006). Customers increasingly prefer localized product and service solutions where content faces adaptation needs for language, images, messages, and positioning. The process for communicating and collaborating within the global eco system of employees, partners, and customers is challenged by cultural differences. Strategic context, content, and process are being re-defined for enhancing communication and learning within the global innovation economy.

The worldwide movement of globalization, innovation, and personalization is driving the focus on knowledge and relationships. The ability to respond to consumer preferences and reconfigure resources dynamically requires a flexible network. A firm's competitiveness is dependent upon its ability to develop a dynamic capability or difficult to imitate combination of resources which includes coordination of inter-organizational relationships (Teece et al. 1997, Eisenhardt and Martin 2000). As international markets demand the design and delivery of localized products and services, organizations are facing increased pressure to optimize knowledge and innovate across the organization.

The development and execution of global innovation strategies are often complex and demanding. There's the continuous quest for capturing scale economies through cost reduction in the value chain as well as scope economies of shared learning across the organization. Organizations have the opportunity to optimize locations that support key activities in the value chain, including R&D, production, marketing and sales, and customer service. Yet organizational infrastructure, technology and social architecture, and human resources influence management of the value chain within the international eco system.

A classic framework that shows the influence of organizational and market drivers on international innovation strategies is the Integration-Responsiveness framework (Doz et al. 1981) which is based upon the needs for global integration of firm resources and local responsiveness to markets worldwide. There are four internationalization strategies that impact local market introductions of new concepts: Responding to pressures for local responsiveness, the home replication strategy or *international strategy* views international business as separate from domestic business, where concepts are designed with local customers in mind and international business is viewed as an extension of the product life cycle. The *multidomestic* or *multilocal* strategy allows more autonomy to the country manager with variance between product and management practices by country (Bartlett and Ghoshal 2002, Cavusgil et al. 2008).

Responding to pressures for more global integration, the *global strategy* approach is where headquarters seeks substantial control over its country operations in order to achieve maximum efficiency, learning, and integration worldwide (Cavusgil et al. 2008). Global integration requires faster and more extensive collaboration with rapid reconfiguration of new opportunities throughout the value chain. While there's the benefit of converging customer needs and the scale of a global marketplace, it's also more challenging to coordinate activities and communication between headquarters and subsidiaries. A more integrative approach for balancing global and local needs is the *transnational strategy* where the firm strives to be more responsive to local needs while retaining central control of operations to ensure efficiency and learning (Ibid). This strategy tries to balance global and local needs through local responsiveness and flexibility while facilitating global learning and knowledge transfer.

Moving from a global standardized approach to a transnational focus, organizations have experienced mixed success in combining a global strategy with a local touch. The creation and application of new business models require internally consistent choices in the areas of customer definition, identification of customer preferences and design for the value creation process. The specific interactions and communication activities that assist firms in accelerating international market responsiveness serve as key connectors in the global value chain.

The highly competitive smart phone market provides interesting examples of global and transnational strategies at work. The top players with the largest share in the worldwide market are Samsung at number one and Apple at number two positions (IDC 2016). Apple enjoys stronger leadership in mature markets such as North America and Europe whereas Samsung holds stronger leadership in emerging markets, especially in Southeast Asia, the Middle East, and Africa. Apple has relied on a global strategy with standardized products and marketing campaigns while localization has been limited to carriers, languages, and portfolio focus. With a strong reputation for radical innovation, Apple has uniquely focused on a universal appeal through product differentiation.

Internally, this translates to a centralized approach where creation and strategic planning phases occur with cross-functional teams based at headquarters. Local teams in key markets are involved in the execution phase for marketing and sales activities which has resulted in standardized products with limited localization and greater focus on the retail store experience in key regions. Although Apple's global strategy has been quite successful since its start, competition has accelerated in emerging markets such as China and India where customers have challenged the lack of attention to local market needs and customer preferences.

On the other hand, Samsung has employed a transnational strategy with a global identity and message while ensuring internationalization and localization when needed in key markets. There is great attention to local preferences through a worldwide team of researchers that are scanning regional market trends and customer needs, resulting in a broad product portfolio for international markets. Samsung adapts products and marketing to local needs and interests with flexibility for local collaboration involving partners and customers. It also offers an open platform for personalization of customer applications.

Inside the organization, this translates to a mix of centralization and decentralization where there is internal collaboration between executive teams, product business units, R&D and subsidiaries to create and plan strategies that shape the Samsung product portfolio and customer experience. The company has a dedicated international team within the Global Strategy Group located at headquarters in Seoul: Activities support business performance and continuous improvement through collaborative innovation and market perspectives on strategic issues. In addition, collaboration between international strategists and senior management contributes to its globalization efforts. The transnational approach with a local focus has helped Samsung accelerate worldwide market growth. However, it has also resulted in the challenge of managing a complex portfolio while ensuring effective communication for validation and decision-making between executives, cross-functional and regional teams.

When exploring the roles of HQ and subsidiaries, both internal and external factors need to be considered. In evaluating the strategic importance of the local market with the resource base of the subsidiary, there is the classic model of subsidiary roles that considers the strategic importance of the market compared to actual resources available for subsidiary performance (Bartlett and Ghoshal 1986). MNC strategy and organizational design influence subsidiary roles where global strategy has an HQ focus and subsidiary dependence, a multidomestic strategy has a decentralized network that is locally responsive, and the transnational strategy has a strategic and local focus in an interdependent network (Harzing 2000). Research on knowledge flows has shown the strategic roles of subsidiaries within the MNC range from global innovators that serve as knowledge sources to local innovators who have complete responsibility of their local market which supports the multi-local strategy. Integrated players receive and give high levels of knowledge flow while implementers rely heavily on knowledge flow from HQ which supports global strategy (Gupta and Govindarajan 1991). Both knowledge and product flows are gaining in importance between subsidiaries which is placing more focus on the transnational strategy (Harzing and Noorderhaven 2006) This provides a view of the organizational structure; however, it doesn't explain the current influence of leadership and team collaboration in a global network.

Internationalizing Radical and Incremental Innovation

The identification of global and local interests helps shape international strategies, where it's also important to pay attention to local market competition and customer preferences that will drive local adaptation (Lemaire 2003). This leads to the question of how firms exploit and explore market opportunities when developing international market strategies. Market discovery is the result of both exploration and exploitation activities where the firm can learn through interactions with teams and customers. When pursuing *exploitation*, the focus is on incremental innovation by exploiting core competencies, defending existing market positions, and maintaining current processes while *exploration* places emphasis on radical innovation by developing new knowledge, entering new markets, and finding new approaches. Incremental innovation is often focused on upgrades or improvements to existing products, services, or programs in existing and new markets, whereas radical innovation is primarily focused on entirely new concepts for existing or new market segments and geographies.

The challenge for organizations is the capability to balance exploitation with exploration in a constantly evolving environment shaped by international market dynamics. In building a strategic picture, the Blue Ocean Strategy outlines a valuable perspective on strategic innovation where those who conform to industry will find themselves in a continuous exploitation mode; however, those who devote more time to exploration can identify opportunities to provide value without market innovation or create unique value and new market opportunities (Kim and Mauborgne 2015). Moreover, it's important to understand international market opportunities for specific geographies: Is there an opportunity to only pursue an exploitation strategy due to high competition and consumer preferences or is there an unexplored market opportunity? What may be an entry barrier in one country may prove to be an untapped market opportunity in another country or region.

In order to respond to customer and market demands, a majority of multinational firms are balancing both radical and incremental innovation in order to ensure market exploration as well as exploitation (Jensen 2014). While a majority of leaders indicate their organizations' front end activities (creation, planning, validation) are centralized, the strategic direction shows an increasing focus on decentralization and local market responsiveness. There is also a growing movement in Asia towards decentralization for front end innovation activities in order to respond to customer needs (Jensen 2015). In view of the increased attention placed on local market knowledge, collaboration with local team members and managers is becoming more important to the success of new products and services.

In planning and organizing international innovation strategies, companies have to consider the roles of headquarters and subsidiaries in terms of communication and knowledge-sharing needs throughout the global innovation cycle. Starting with divergent communication flows to gather ideas and knowledge from around the world, followed by convergent communication flows to bring knowledge together for synthesis and decision-making. The challenge for leaders and management teams is the ability to evaluate multiple market needs while ensuring sufficient knowledge-sharing from subsidiary teams. When introducing new products and services to international markets, there's an interdependent process between HQ and subsidiaries for planning and execution. The organization's ability to recombine and reconfigure local market knowledge from subsidiaries influences its ability to create, develop, market and sell solutions around the world.

The strategic context serves an influential role in shaping the development of knowledge and capabilities for innovation. When introducing new concepts to international markets, the project leader and multicultural teams apply global and local perspectives that are integrated through organizational routines. Social embeddedness and relations become essential to the development of strategy and capabilities through shared understanding and interactions in strategy-making (Regner and Zander 2011). Shared understanding in strategy-making for global innovation becomes an important focus due to the complexity of introducing new concepts to diverse markets, cultures, and customers.

Optimizing Creativity through Cultural Diversity

New concepts often require an integrated approach with a front-end process followed by a rigorous project management process for new product development (NPD) and go-to-market (GTM) stages. There are new product introductions that may require both radical and incremental innovation depending on the strategic vision, the market demands, and related technologies. This requires a strategy-making process involving communication and social interactions for conceiving and bringing new concepts to international markets. When considering innovation systems as social systems, there is a process of "social making" of innovations that can define a socially accepted space determined by cultural interactions including: affective frames of identity and difference, cognitive frames of knowledge and normative sets of values, norms, and beliefs (Pohlmann et al. 2005). A safe and open space where cross-cultural teams can share and collaborate is essential to strategic co-creation.

Creativity as a team practice and competency is drawing attention from organizations as they seek to strengthen global innovation efforts. Arts-based and design thinking methods are driving this momentum since they rely on diverse perspectives and a customer-centric view compared to traditional data analysis and problem-solving methods. With the growing focus on innovation, arts-based methods have attracted more interest from organizations who are seeking to better prepare their managers for a dynamic market (Adler 2006). In offering a universal language, art-making and visual thinking are becoming effective methods for unifying diverse perspectives and creating new meaning. Leaders and managers can use art to facilitate sense-making across cultures (Grisham 2006). Arts-based methods involve four key elements in the creativity process which are found in various art forms: Skills transfer, projective techniques, illustration of essence, and making (Taylor and Ladkin 2009). In a cross-cultural context, arts-based methods are especially useful in bridging cultural gaps in order to facilitate sense-making and connection within a global team.

Design thinking is viewed as a "human-centered" discipline that involves the application of traditional designer skills for identifying problems and inventing solutions with multidisciplinary teams, their clients, and the users

(Brown 2008). Using a three-phase process of inspiration, ideation, and implementation, design thinking activities are often applied within diverse teams for encouraging creative collaboration. This requires empathy, integrative thinking, optimism, experimentation, and collaboration (Ibid). Arts-based and design thinking methods can facilitate multicultural collaboration for creation and planning phases, however team leaders and facilitators need to consider techniques and tools that support dialogue for culturally diverse communication and knowledge-sharing practices.

The key to success for international concepts is the front end innovation process which involves the concept creation, strategic planning and validation phases. This exploration stage combines insights, ideation, opportunity analysis, concept validation and strategy-making of new products which requires effective collaboration with cross-cultural and cross-functional teams. The Front End Innovation (FEI), New Product Development (NPD), and Go-to-Market (GTM) stages are interdependent and essential to managing the planning and execution of new product and service concepts.

The Challenges of Global Strategic Planning

When examining critical knowledge required for the planning and execution phases of the global innovation project, the strategic planning phase is identified as the critical point of interaction between global project leaders and management teams in HQ and the local teams based in subsidiaries (Jensen 2014). Local market, customer, and product knowledge is sought by the global project leader where customer validation, resource allocation, and local product feature needs are critical for planning. The planning phase relies on knowledge of local market requirements to determine the level of standardization or adaptation required for a new concept. Country knowledge concerning customer preferences, user practices, competition, and market trends are essential during this phase where local managers can provide insights to customers and markets. Involvement of local teams, customers, and partners who can validate and support the offer is also important in the early project phases in order to ensure a common vision and strategic alignment.

When exploring the role of local team members in subsidiaries during the front-end innovation process, a majority are involved in the tactical details of launch preparation and go-to-market implementation (Jensen 2014). Some organizations ensure earlier involvement at the validation phase to ensure local product adaptation while a fewer number allow participation between HQ and local teams for ideation and planning. The ideation and planning functions are often viewed as roles for HQ management in evaluating and organizing global market needs and opportunities. In reviewing information sought by global project leaders for the planning phase, there's a strong focus on local market, customer, and product knowledge in order to understand market potential and customer preferences (Jensen 2014). The nature of this information indicates the planning phase places a strong emphasis on access to and sharing of local market knowledge. Access to local market and customer knowledge is often made possible through specific interactions with local team members.

The main point of conflict that exists between the global management team based in HQ and the local teams is the perception and understanding of global and local team knowledge in conceiving and bringing new products to market. The project collaboration process primarily involves centralized planning at HQ driven by the global project leader and decentralized execution driven by local team members in key markets (Jensen 2011, 2014). The global team leader in HQ is driving centralized planning, ideation, and validation processes without or with variable participation by local team members. The lack of knowledge-sharing during the ideation and planning process prevents or limits local team members from contributing their cultural and market knowledge. The lack of access to local market intelligence often results in new concepts and products that are poorly adapted to international market and customer needs.

As shown in Fig. 3.1, local team members usually do not have the opportunity to contribute to the creation of a new concept or solution during the front-end innovation process. Local team participation during the global innovation cycle is primarily focused on execution, where global project leaders reported that only a small percentage of local teams are involved in ideation (11%) and strategic planning (16%) with slightly higher participation (22%) during the validation phase (Jensen 2014). The lack of participation in the front end combined with an emphasis on execution may contribute to reduced interest and motivation on two levels: (1) contribution to the creation of new concepts for future product introductions and (2) marketing and selling the new concept. In view of the tensions between global strategy and international market execution, local engagement with teams and customers are critical activities for improving market responsiveness.

Under pressure to meet business performance objectives, cross-cultural teams need to meet expectations for increasing global and local sales revenue and market share, customer adoption and satisfaction, as well as product quality and performance (Jensen 2011, 2014). This means a continuous

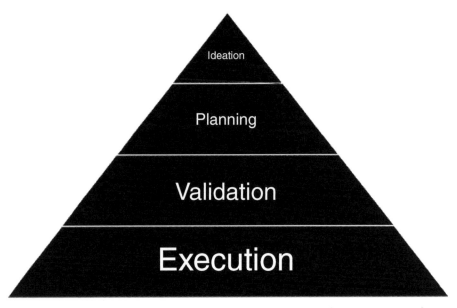

Limited participation in front-end innovation with local teams focused on execution.

Fig. 3.1 Local participation in global innovation cycle phases

drive to manage cost and budget objectives, product and service quality, market positioning and differentiation, as well as timely delivery to market. It's important to mitigate these pressures and related stress across cultures and within the organization. In order to ensure team and project performance, leaders emphasize the importance of effective cross-cultural interactions, communication, and collaboration skills. However, only 31% of team leaders noted that global team collaboration and knowledge-sharing were part of project performance measures (Jensen 2014). The primary focus for team performance is still driven by the project process and business results, including time to market, product quality, customer adoption, and sales.

> **Nokia's Strategic Challenge: Moving from Global Implementer to Local Collaborator**
>
> Nokia was a leading global manufacturer of mobile devices with a broad product portfolio for business and consumers in mature and emerging markets. Controlling 41% of worldwide market share in 2007, the company enjoyed a strong reputation for innovation and performance in mobile devices. However, Nokia faced great challenges between 2008 and 2013 when it failed to recognize and respond to the smartphone revolution during a period with rapid changes in technologies and user preferences.

In order to remain competitive and innovative, Nokia welcomed new leadership from CEO Stephen Elop of Microsoft in 2010 who brought operational expertise and supported major organizational changes designed to increase collaboration, innovation, and execution efficiency. The organization needed to accelerate the development and introduction of its line of Lumia smartphones. While Apple, Samsung, and other competitors were increasing international sales of smart phones, Nokia experienced challenges in maintaining its market lead for product innovation and responsiveness to consumers in some of its key international markets.

There have been several perspectives offered by analysts, researchers, and the press on the reasons for Nokia's demise in the global smartphone market. Conclusions included being too comfortable with its leadership position and global strategy, lacking urgency and the inability to take risk, a top-down and execution-focused leadership style with lack of transparency, and ignorance of the US market. One could conclude that leadership, strategy, the organizational culture and team climate were not completely aligned within the organization. Nokia had experienced past market success through a transnational strategy where concept development and strategic decisions for front end innovation were centralized at HQ yet adaptation and localization occurred where needed, especially for emerging markets. This created a knowledge-sharing structure where local teams served as contributors and implementers with some participation in planning and validation phases, and full participation for execution. Its focus on execution efficiency became a weakness due to the lack of knowledge-sharing and ownership for local teams. There was limited opportunity and time for international teams to pursue creativity and innovation, propose ideas and take new initiatives. The organizational matrix structure with complex reporting and decision processes complicated the ability to openly communicate and share knowledge.

Under new leadership, the company experienced rapid organizational change to improve responsiveness to local markets. In order to increase local market responsiveness and collaboration, Nokia moved to a strategic co-creation process with subsidiaries worldwide. In order to improve its focus on execution efficiency, Nokia placed a stronger emphasis on local market responsiveness and global team transparency in order to increase ownership and accountability for team members. As noted by a former program director at the Chinese subsidiary: "We need to share and let local teams feel the responsibility. Accountability is welcomed. We are now asking for more and receiving more". Local team members participated in front-end innovation which resulted in intrapreneurial and collaborator roles and an acceleration of new ideas through increased communication with local markets.

Local engagement through strategic co-creation was introduced at a very late stage of the company's change initiative. As a result, Nokia was not able to sufficiently accelerate organizational innovation for increasing international market share of its smart phone offering. It could not reverse declining sales and increasing competition, and eventually sold its devices and services business to Microsoft in 2014 in order to focus on its networking business. However, Nokia has recently announced a return to smartphones in 2017. Stay tuned, as lessons learned in past years may transform into new innovations and international market opportunities.

> Sources: Cheng, Roger. "Farewell Nokia: The Rise and Fall of a Mobile Pioneer", April 25, 2014: http://www.cnet.com/news/farewell-nokia-the-rise-and-fall-ofa-mobile-pioneer/.
>
> Jensen, Karina R. Local Innovation and Collaboration Study/Asia, 2015.
>
> Oakley, Phil. "Nokia has announced a smartphone comeback in 2017 – powered by Android", Business Insider UK, December 1, 2016: http://uk.businessinsider.com/nokia-has-announced-a-smartphone-comeback-in-2017-powered-by-android-2016-12?r=US&IR=T.
>
> Schrage, Michael. "The Real Cause of Nokia's Crisis", Harvard Business Review, February 15, 2011: http://hbr.org/2011/02/the-real-cause-ofnokias-crisi.html.
>
> Huy, Quy and Vuori, Timo. "Who Killed Nokia? Nokia Did", INSEAD Knowledge, September 22, 2015: http://knowledge.insead.edu/strategy/who-killed-nokia-nokia-did-4268.

Local Roles within an Interdependent Organizational Network

"We have such a broad product and services portfolio with business units creating their own product lines," exclaimed a global marketing executive from a US based multinational firm in the high-tech industry, adding with a frustrated tone "the geographies have a challenge in responding to this demand, knowing what to communicate and how to communicate, what data with what purpose". Working in a fast-paced environment with aggressive competition and continuously shifting customer preferences in international markets, the executive felt the constant pressure to clarify strategic focus for teams worldwide. The heavy focus on execution did not allow for sufficient time on strategic planning and the integration of new ideas and knowledge from the regions.

The role of local management teams in subsidiaries determines the commitment and capability to introduce new concepts in local markets. Due to the interdependent nature of conceiving and introducing new concepts, HQ and subsidiary teams hold critical and complementary knowledge that need to be integrated for successful execution. Yet organizational knowledge flow appears to move primarily from global headquarters to subsidiaries during the strategic planning phase where most communication is either initiated at global headquarters with feedback from local subsidiaries (60%) or directed from global headquarters to local subsidiaries (31%). There is less communication at the local level (9%) including communication initiated from subsidiaries to global headquarters and communication conducted between local subsidiaries (Jensen 2014). Knowledge flow between headquarters and subsidiaries can effectively influence collaboration between global and local teams.

The opportunity for local managers to initiate and communicate local market and cultural knowledge contributes valuable information for concept creation and strategic planning activities, in addition to effective execution of products and services. Research has shown the importance of decentralization, subsidiary management credibility, communication, and a global perspective in determining entrepreneurial initiatives at the subsidiary level (Birkinshaw and Hood 2001; Verbeke 2009). When addressing internal collaboration and HQ attention, subsidiaries with a high level of strategic choice and value adding activity perform better (Ambos and Birkinshaw 2010). The ability to respond to local market opportunities requires an effective knowledge-sharing process for geographically distributed teams within the global network.

In order to better understand the motivations and challenges that can influence knowledge-sharing and contribution from local team members during the innovation process, a comparison was made between the motivations, challenges, and critical incidents from the Global Innovation and Collaboration study (2014). This evaluation resulted in a knowledge-sharing structure with four distinct project collaboration roles served by local subsidiary teams during the global innovation project cycle. They are based on the descriptions from global project leaders and local team managers concerning challenges and motivations for local team members. As shown in Fig. 3.2, the four roles

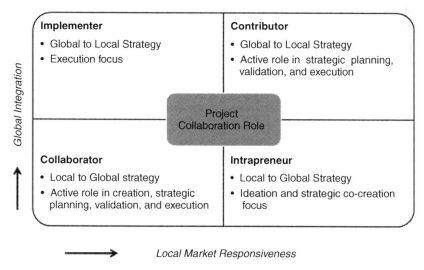

Fig. 3.2 Local team roles for the global innovation project

identified include: **Implementer, Contributor, Collaborator** and **Intrapreneur**. These roles were then mapped to the international innovation strategy in order to evaluate the emphasis on global integration (global market exploitation and integration) versus local responsiveness (local market exploration and responsiveness).

When examining roles more common to global integration and exploitation, the *Contributor* and *Implementer* roles are more involved in the execution phase with limited or no participation in ideation, planning, and validation. The Implementer role is identified with an emphasis on execution and support of a global strategy due to top-down planning where there is no involvement in the front-end innovation process. This role appears to create challenges for knowledge-sharing and collaboration where there's a lack of motivation for local team members. Supporting a global to local strategy, the Contributor is expected to provide more feedback and suggestions for adapting global concepts to local markets. While there is no opportunity for ideation and initiative, there's more motivation linked to the opportunity to participate in validation and adaptation activities for local markets.

The roles of *Collaborator* and *Intrapreneur* are focused on local to global co-creation where there appear to be higher motivations for knowledge-sharing and collaboration in the front end. The Collaborator drives local to global strategy by sharing local market knowledge and understanding how to integrate global and local market opportunities. The Intrapreneur initiates and proposes new ideas and concepts that contribute to local market strategies as well as global business objectives. There's an increased focus on local co-creation and multicultural collaboration. In the words of an international product management director: "When their feedback is taken into account, the local market feels you've created something for them and they're willing to create a great GTM program and accelerate sales."

In evaluating challenges for knowledge-sharing and collaboration, the role of the *Implementer* emerged from an emphasis on execution and support of global strategy during the project collaboration process. In describing the launch project process, local team members referred to the daily pressures of executing and selling new products into local markets (Jensen 2014). There is not sufficient time to focus on future market opportunities and share knowledge since there are ongoing global initiatives to support.

Local team managers in Asia noted the emphasis on top-down planning for global strategies where the global project leader and top management in HQ determine the conception and direction for new concepts (Jensen 2015). Local team members are expected to support the global strategy and do not have the opportunity to share their knowledge. This leads to the feeling that

HQ management does not have an interest nor an understanding of local market opportunities as well as local customer knowledge. Thus, the implementer role is expected to focus on the execution phase without involvement in the ideation, planning, and validation phases.

Supporting and contributing to the global strategy, the role of the *Contributor* is primarily based on the ability of teams to provide feedback and suggestions for global concepts. The knowledge-sharing role of local team members is recognized by the global project leader and the HQ team since local market validation and planning is needed in order to ensure successful execution. Local team members provide knowledge on local adaptation needs for introducing global concepts where concept creation and planning have already been determined with top management teams in HQ. However, the need for more engagement in contributing and providing feedback on global concepts was expressed by local study participants as well as global project leaders. This includes the opportunity to actively participate in the creation and strategic planning phases as well as ensuring open communication and knowledge-sharing within the MNC network.

The role of the *Collaborator* is to shape local to global strategy by sharing local market knowledge and understanding how to integrate global and local market opportunities. This role is preferred by a majority of local study participants as well as global project leaders in order to improve interactions between the HQ and subsidiary teams (Jensen 2014). The global project leader and the management team in HQ recognize and respond to the knowledge-sharing practices of local team members. There is an interest by the collaborator to create relationships and increase knowledge-sharing within the MNC network. This involves travel, visits and exchanges by local team members in order to increase interactions and to be fully integrated in the global network. The collaborator role promotes transparency and open communication concerning local market practices. This means active participation in the creation, strategic planning, validation, and execution phases. The collaborator role also has an awareness of the necessity to structure and present communication that is easily understood by global team members in order to ensure effective knowledge-sharing.

Finally, there is the role of the *Intrapreneur* who initiates and proposes new ideas and concepts that contribute to local market strategies as well as global business objectives. The local manager has full recognition and support from the management team in HQ to pursue new initiatives. This role appeals to local managers that are ready to take on ownership of a new project and feel empowered by the recognition and rewards available from the global MNC network. The role emphasizes active participation in the creation and strategic planning phases with variable involvement in validation and execution phases depending on the responsibilities of the local manager. Local managers that take on an intrapreneurial role appear to be very engaged in the local and global innovation process in order to create an impact within the organization. There is a strong interest and ambition in taking on more responsibilities and contributing to new ideas and products that make a local and global impact. This requires an organizational culture and leadership that promote initiative, risk-taking, creativity, and collaboration throughout the MNC network.

From Strategic Planning to Strategic Co-creation

When creating new concepts for international markets, there's always the challenge for the management team in HQ to capture accurate information on planning needs such as pricing, features, and customer preferences. On the other hand, subsidiaries have a difficult time to understand the global and strategic view from HQ. For example, a telecommunications firm based in Europe faced increased pressure to accelerate performance for key international markets due to global competition. Local teams were especially frustrated with the current collaboration process since it did not allow effective feedback. As noted by the senior manager of global products "It's important to communicate the big picture for teams…you can't give the impression that HQ directs everything. It's important for the local team to understand needs and interests."

The firm saw great improvements when it decided to organize trips around "global information sessions" where the executives and product management teams could meet with local teams from various subsidiaries worldwide. At the start of each year, they organize a worldwide tour to meet local sales teams in order to listen to their challenges. The senior manager is very enthusiastic about the results noting that "We ensure executives and product managers can travel to various sites and unite local teams. This helps communication and learning about products."

A shared vision and strategic co-creation can influence local engagement and international market results. There is a growing movement from centralized to decentralized decision-making in order to focus on co-creation and strategy-making. Due to the need for increased local market knowledge and collaboration, organizations can benefit from the practice of strategic co-creation (see Fig. 3.3). This requires leadership that can ensure recognition and responsiveness to new ideas and knowledge shared by local teams. Engagement of local teams in the concept creation and planning phases through meaningful contribution can increase motivation as well as project success. It also provides opportunities to invite co-creation with local customers and partners.

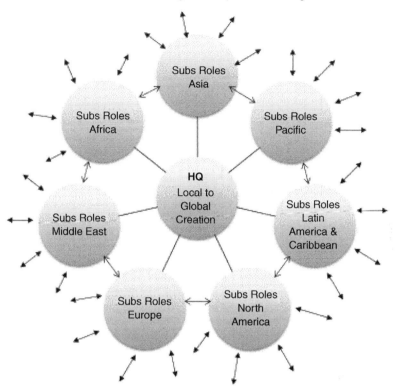

Fig. 3.3 Strategic co-creation through local to global integration

An innovation strategy that is focused on a local to global strategy-making process emphasizes decentralization and local market responsiveness. When developing the global innovation strategy, senior management needs to engage the local teams as planning partners in order to create a shared understanding of strategy-making. During the front-end innovation process, the concept creation and strategic planning phases are co-created at the local level for integration at the global level. This allows organizations to move from global strategy to strategic co-creation.

In order to increase success in conceiving and executing innovation strategies for international markets, multicultural collaboration should serve as a competitive advantage and critical resource for MNCs in accelerating innovation and market responsiveness. Cross-cultural team interactions facilitate the sharing of local market knowledge, cultural understanding, and the creation of new ideas. Increased collaboration can be achieved through a focus on knowledge-sharing and participation in the front end of innovation, specifically the creation, planning, and validation phases (Jensen 2014). The orchestration and reconfiguration of organizational resources combined with project collaboration routines create front-end innovation process capabilities.

> ### Philips Focus on Global Innovation and Design through Co-Creation
>
> Philips is a world leader in healthcare, consumer lifestyle, and lighting products that has successfully applied global innovation as a key driver for market growth. It offers a broad product portfolio to consumers and businesses with global and local solutions for mature and emerging markets. Its vision is to make the world healthier and more sustainable through innovation, creating value through better health and better care at lower costs while delivering energy-efficient and connected digital lighting.
>
> Since 2011, Philips has been firmly focused on the Accelerate! Journey of change and performance improvement. It is designed to transform the company into an agile and entrepreneurial mindset in order to deliver valuable, effective, and timely innovations to customers in local markets. It shapes the organizational culture values and behaviors around three themes – Eager to win, Take ownership, and Team up to excel. The Path to Value is based on three drivers: Initiating new growth engines, expanding global leadership positions, and transforming or addressing underperformance.
>
> The leadership challenge is to inspire multicultural team collaboration and market innovations that continue to meet expectations of customers and markets around the world. In a constantly evolving global business environment with changing customer preferences, there is an increased need for market responsiveness. Local teams require sufficient time and space to share knowledge about customers and market opportunities. In order to align cross-cultural and cross-functional teams around a common goal, there is a need for transparency and trust-building for sharing critical knowledge that leads to new market opportunities.
>
> Through Philips focus on co-creation, design has become an integrated strategic function that has resulted in cross-functional and cross-cultural collaboration across business units and geographies. It's a shared responsibility and process that is developed through people-centric and customer-centric design and innovation. Philips Design Strategy and Design Innovation group drives strategic planning, performance management, and business improvement. The Design Research and Innovation program creates new design competencies, future visions, and new propositions for creating meaningful and valuable solutions. Design thinking serves an important role in experimentation, collaboration, and development of new solutions that bring a holistic view of the end user and the experience.
>
> Driven by insights and understanding of diverse cultures and customers, solutions are co-created within ecosystems that link user needs. The co-creation process involves stakeholders and customers where they explore the business environment and the experience of interacting with products and services. Multidisciplinary teams have the opportunity to understand user needs from the user perspective that transform into ideation, creating and testing new concepts.
>
> Front-end innovation activities are decentralized while integrating radical and incremental innovation that incorporates the latest technologies. With

regional headquarters, the company structure represents an integrated global network that does not have any physical boundaries which facilitates team collaboration, from concept creation to execution. The global project lead could be based in the US while the technology is developed in Europe and markets are located in Asia. In addition to using communication technologies to bridge geographic distances, global meetings are held at key geographic locations in order to engage regional and local team members in ideation and planning that provide new insights to the product portfolio.

The company empowers multicultural team collaboration where leaders have global responsibility regardless of their location. Leadership is about building trust and relationships that strengthen local connections and create an inclusive global community. Leaders become facilitators that enable teams to bring new innovations through collective intelligence, optimizing ideas and knowledge from teams worldwide. In order to foster cross-cultural dialogue, there is an emphasis on collaborative spaces including Creative Studios and Innovation Hubs where teams can co-create. "We have designed specific co-creation spaces that allow for movement and flexible layouts which support rapid ideation," noted Kurt Ward, Senior Design Director, when describing co-creation sessions for leaders and teams around the world.

Philips ensures full participation of cross-cultural and cross-functional teams involved in the creation and introduction of new concepts that deliver value for customers in local markets. Through co-creation and global transparency, teams can ensure increased market responsiveness and execution efficiency. Open and interactive team collaboration is strengthened through a shared design methodology and tools. Project leaders facilitate knowledge-sharing and collaboration through inclusive and empowering leadership roles. With the opportunity to take initiative and fully engage in front-end innovation, teams are inspired to create and deliver customer-centric solutions that improve people's lives throughout the world.

Sources: Company page, Philips: http://www.philips.com/a-w/about/company.html. Accessed on November 1, 2016.

Interview, Kurt Ward, Senior Design Director, Philips Design, November 25, 2016.

Jensen, Karina R. Global Innovation and Collaboration Study, 2014.

Jensen, Karina R. Local Innovation and Collaboration Study/Asia, 2015.

Our Design Manifesto. Philips Design page: https://www.90yearsofdesign.philips.com/manifesto. Accessed on November 1, 2016.

Strategic Co-creation Audit

On a whiteboard, electronic board or your laptop/tablet, use a world map to identify countries and their strategic roles:

1. Which countries are viewed as strategic markets?
2. Which countries are viewed as high potential markets?

3. What is the participation status of local team members for global innovation phases?
   - Creation/Ideation
   - Strategic Planning
   - Validation
   - Execution

4. What are current roles of local team managers?
   - Intrapreneurs
   - Collaborators
   - Contributors
   - Implementers

5. Review local participation and local team roles at present.
6. Identify desired participation and roles of local team managers in strategic and high potential markets.
7. Compare actual roles with desired roles.
8. Consider opportunities for team engagement in the planning and execution of the global innovation strategy.

Bibliography

Adler, Nancy J. "The Arts and Leadership: Now that We Can Do Anything, What Will We Do?", Academy of Management Learning and Education, 5–4 (2006): 486–99.

Ambos, Tina C. and Birkinshaw, Julian. "Headquarters Attention and Its Effect on Subsidiary Performance", *Management International Review*, 50(2010): 449–69.

Apple info page, Apple: http://www.apple.com/about/. Accessed on January 10, 2017.

Apple Reports Fourth Quarter Results, Apple: http://www.apple.com/newsroom/2016/10/apple-reports-fourth-quarter-results.html. Accessed on January 10, 2017.

Bartlett, C.A. and Ghoshal, S. *Managing Across Borders: The Transnational Solution*. 2nd ed. Boston: Harvard Business School Press, 1986, 2002.

Birkinshaw, J. and Hood N. "Unleash Innovation in Foreign Subsidiaries," *Harvard Business Review*, 79–3 (2001): 131–37.

Bradshaw, Tim. "Apple Faces China Challenges Despite IPhone Resurgence", Financial Times, February 1, 2017: https://www.ft.com/content/474aec32-e82e-11e6-893c-082c54a7f539.

Brown, Tim. "Design Thinking", *Harvard Business Review*, 86–6 (2008): 84–92.

Cavusgil, S. Tamer, Knight, Gary and Riesenberger, John R. *International Business: Strategy, Management, and the New Realities*. New Jersey: Pearson Prentice Hall, 2008.

Davenport, Thomas H., Leibold, Marius and Voelpel, Sven. *Strategic Management in the Innovation Economy: Strategy Approaches and Tools for Dynamic Innovation Capabilities*. Erlangen, Germany: Publicis and Wiley, 2006.

Doz, Yves L., Bartlett, Christopher A. and Prahalad, C.K. "Global Competitive Pressures and Host Country Demands: Managing Tensions in MNCs". *Sloan Management Review*. (Spring 1981): 63–74.

Eisenhardt, Kathleen M. and Martin, Jeffrey A. "Dynamic Capabilities: What Are They?" *Strategic Management Journal*, 21:10–11 (2000): 1105–21.

Global Strategy Group, Samsung: https://sgsg.samsung.com/main/newpage.php?f_id=gsg_whyGsg. Accessed on December 11, 2016.

Grisham, Thomas. "Metaphor, Poetry, Storytelling and Cross-Cultural Leadership", *Management Decision*, (2006): 44–4; 486–503.

Grobart, Sam. "How Samsung Became the World's No.1 Smartphone Maker", Bloomberg, March 28, 2013

Gupta, A.K. and Govindarajan V. "Knowledge Flows and the Structure of Control within Multinational Corporations", The Academy of Management Review, 13/4 (1991): 768–92.

Harzing, AW. "An Empirical Analysis and Extension of the Bartlett and Ghoshal Typology of Multinational Companies", *Journal of International Business Studies*, 31(2000): 101–120.

Harzing, Ann and Noorderhaven, Niels. "Knowledge Flows in MNCs: An Empirical Test and Extension of Gupta and Govindarajan's typology of subsidiary roles", *International Business Review*, 15–3 (2006): 195–214.

Huy, Quy and Vuori, Timo. "Who Killed Nokia? Nokia Did", INSEAD Knowledge, September 22, 2015: http://knowledge.insead.edu/strategy/who-killed-nokia-nokia-did-4268.

IDC. Smartphone Vendor Market Share 2016 Q2. Accessed on December 11 2016, http://www.idc.com/prodserv/smartphone-market-share.jsp.

Jensen, Karina R. Accelerating Global Product Innovation through Cross-cultural Collaboration Study, Report 2011.

Jensen, Karina R. Global Innovation and Collaboration Study, 2014.

Jensen, Karina R. Local Innovation and Collaboration Study/Asia, 2015.

Kim, W. Chan and Mauborgne, Renée. *Blue Ocean Strategy*. Boston: Harvard Business School Publishing, 2015.

Lemaire, Jean Paul. *Strategies d'Internationalisation: Développement International de l'Entreprise*, Paris: Dunod, 2003.

Pohlmann, Markus; Gebhardt, Christiane and Etzkowitz, Henry. "The Development of Innovation Systems and the Art of Innovation Management – Strategy, Control, and the Culture of Innovation", Technology Analysis & Strategic Management, 17 (2005): 1–7.

Regner, Patrick and Zander, Udo. "Knowledge and Strategy Creation in Multinational Companies," *Management International Review*, 51 (2011): 821–50.

Reuters. "Here's Why India Will Be Such a Challenging Market for Apple", May 24, 2016. http://fortune.com/2016/05/24/india-market-apple-tim-cook/.

Taylor, Steven S. and Ladkin, Donna. "Understanding Arts-based Methods in Managerial Development", *Academy of Management Learning and Education*, 8–1 (2009): 55–59.

Teece, David J., Pisano G., and Shuen, A. Dynamic Capabilities and Strategic Management, *Strategic Management Journal*, 18:7 (1997): 509–33.

Values and Philosophy, Samsung: https://sgsg.samsung.com/main/newpage.php?f_id=samsung_value. Accessed on December 11, 2016.

Verbeke, Alain. *International Business Strategy: Re-thinking the Foundations of Global Corporate Success*. Cambridge: Cambridge University Press, 2009.

Part II

Dialogue: Nurturing Knowledge-Sharing and Learning

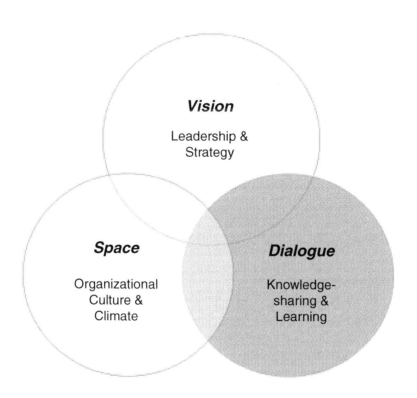

4
Communicating in a Multicultural and Networked World

A seasoned global product director working for a US-based high tech multinational firm enjoyed both the challenges and opportunities in facilitating collaboration amongst his cross-functional and cross-cultural teams. With work experience and cultural knowledge across regions, in addition to assignments in both Asia and North America, he was especially intrigued by the differences in Eastern and Western cultures during the innovation process. In working with local teams, he noted "If I go to Shanghai as a team leader and I ask the local team to brainstorm and how to solve ideas, there will be silence and nobody will say anything. They will instead say "you are the leader, tell us what to do, how to do it for you?" It is more common in China for the leader to tell the idea and show the process. There is also the challenge and fear of sharing should an idea be considered "wrong" or cause conflict in that it is too different from other ideas shared. The director emphasized that "it may be easier to get ideas through one-on-one sharing or in a group environment where the boss is not in the room so they can speak more freely".

The approach to sharing ideas and knowledge contrasts greatly between East and West, especially in comparing Chinese and American perspectives. "Looking back, I have seen a few cases where Chinese nationals feel that seniority allows more authority and that others should follow what they do and say, although they may not have all of the knowledge. American team members may not agree since open dialogue is preferred and by sharing more feel they have more knowledge and influence," the director paused and then added with a bemused look,

"The challenge is that neither the Chinese or American team member is open to other cultural perspectives, the disagreements may escalate and drag down the entire team during the innovation process."

The Role of Culture in Global Innovation

The role of multicultural collaboration within the context of global innovation has received limited attention in both research and practice. Yet the greatest challenges in successful execution of global innovation projects involve effective communication and knowledge-sharing with teams worldwide (Jensen 2014). Organizations are struggling to be more responsive and to connect with local customers and markets through improved communication and knowledge-sharing between global and local teams. Increased interactions for identifying, designing, and delivering new solutions require improved awareness and understanding of cultural communication differences and similarities.

Communication becomes more challenging with distance, culture, language, and time zone differences for geographically distributed teams. Cultural perspectives for knowledge-sharing need to be considered from North to South, West to East. The arrival of social media and new technologies has created both challenges and opportunities for optimizing cross-cultural knowledge within the global network. The structure and flow of communication needs to be carefully organized in order to ensure sufficient access, understanding, contribution, and ability to share knowledge. It becomes necessary to promote dialogue through interpersonal and virtual communications. In order to fully prepare for international collaboration, there is an increasing emphasis on developing the cross-cultural competencies of leaders and teams around the world.

The Dynamics of Cross-Cultural Communication

A point of tension in leading and managing cross-cultural teams is the delicate balance between relationship development and project execution. As a leader, you need to ensure sufficient time for developing trust through relationship-building, knowledge-sharing, and learning in order to ensure successful team collaboration. In addition, there needs to be a continuous focus on team roles, engagement and delivery on project objectives. This exemplifies Hall's (1976) high context and low context model, a classic and still highly relevant model that points to the emphasis on relations in contrast to time and task focus. On one hand, there's the high context culture that values relationships represented by such countries as Japan, China, Brazil – time is needed to build trust prior to the start of a project while communication and socialization are needed throughout the project process. On the other hand, there is the low context culture that emphasizes task and time management represented by such countries as the US, Germany, Sweden and the Netherlands – a focus on project objectives and execution is a priority where relationship-building is expected to develop along with the project.

When applied to multicultural teams there needs to be consideration for socialization and communication activities in order to build trust. Yet there also needs to be a project focus where there is a common vision and agreed upon roles by team members. Leaders have the opportunity to shape a healthy team dynamic where culturally diverse team members can easily interact and manage a balance between relationship-building and project execution. There is also the communication style where High Context cultures may use indirect and implicit language, while Low Context cultures prefer direct and explicit language. The ability of the leader to create a team environment that can facilitate project collaboration and effective execution through direct and indirect dialogue will have an impact on project performance and market success.

Communication is always more challenging when working across geographic and cultural distances. Sources of cross-cultural misinterpretation include subconscious cultural "blinders", a lack of cultural self-awareness, projected similarity, and parochialism (Adler 2007). When faced with cross-cultural interactions, effective businesspeople should assume difference rather than similarity (Ibid). It is generally a good idea to assume there will be differences yet also explore and identify similarities. Cross-cultural communication requires a reciprocal exchange for developing cultural understanding. Each party needs to consider how much they will share from their own cultural perspectives and practices as well as the time and consideration they offer in listening, observing, and learning about other cultures. This results in an inclusive and insightful dialogue that nurtures cultural empathy and understanding.

Language and Project Communication

When facilitating multicultural collaboration, there is always the challenge of multilingual interactions where team members do not share the same native language. Understanding the interplay between the multiple facets of language and how they affect day-to-day operations is becoming important for global business effectiveness (Piekkari et al. 2014). It can add more complexity to the project collaboration process depending on the mastery of foreign languages by team members. Most organizations elect a corporate language in order to sustain the dialogue worldwide, most often this language is English. However, even with a single language, there is the issue of how effectively team members can communicate with each other. Misunderstandings and surprises can occur if team members do not fully master a language, including issues such as incorrect translation of words, pronunciation, and lack of communication due to discomfort for non-native speakers.

When words are lost or mixed in translation, the final message may not always be well received. Travel encounters often result in funny situations when you realize that you ordered the wrong item on the menu or you said goodbye to the host when you meant to say hello in the local language. However, when you're in the midst of an innovation project and you're not sure if the global team really understood the new feature requirements or the new campaign theme, it's time to get an interpreter and check for local language context. Memorizing "hello", "thank you" and "cheers" in 20 different languages can make a good impression for team meetings and celebrations, but don't forget to check for translation when you have critical team and client discussions.

Incorrect use or lack of knowledge in a language can impact communication and knowledge-sharing practices that occur during the global innovation project cycle. The ease or difficulty in understanding the team's dialogue with each other depends on the level of English used, the type of accents and their influence on pronunciations of the English language, in addition to interpretations formed between the sender and receiver. As noted by a senior manager: "Sometimes important meetings are only held for 20 minutes since people don't want to talk too much and they're not comfortable talking. We miss details and we can't replace this with phone calls or emails."

While it's important that everyone has the opportunity to speak and express their thoughts and ideas, actual engagement and communication by team members could be challenged by discomfort in their level of English and the ability to express their ideas. On the other hand, there are members that are at ease with English and take more initiative in sharing their ideas which could result in some team members dominating the conversation over others, losing valuable knowledge and insights. Thus, there is a need for a communication structure and process that enable contribution and engagement from every team member across the organization.

Cultural Differences in Knowledge-Sharing

Knowledge-sharing through team interactions is a core process in facilitating multicultural innovation and collaboration. Knowledge-sharing is defined as "the provision of receipt of task information, know-how and feedback on a product or procedure" (Foss et al. 2010) which is often a crucial antecedent to knowledge creation (Cohen and Levinthal 1990, Tsai 2001, Nonaka 1994). Opportunities for co-creation and collaboration happen when team members provide and receive information, knowledge, and feedback on their experiences with products, services, and customers.

The project collaboration process enables knowledge-sharing through informal and formal interactions in each phase of the global innovation cycle, from concept to design to market. It starts with the connections that are made with local customers and markets that inspire new ideas to be shared and synthesized with cross-functional and cross-cultural teams. A final concept is co-created, designed, and developed that involves continuous communication and sharing within the global network. Finally, the concept is transformed into a solution for customers who inspire another cycle of co-creation and innovation around the world.

Cross-cultural management as a direct link to knowledge management holds great relevance in establishing cultural knowledge as an organizational resource that supports business and operational objectives. The knowledge-based concept of cross-cultural management is based upon a networking behavior for facilitating the transfer of organizational knowledge and experience (Holden 2002). In this way, multicultural collaboration provides a competitive advantage in facilitating knowledge that responds to local market opportunities.

Since knowledge-sharing plays such a key role in global innovation and collaboration, it is surprising that few studies have explored cultural differences. The research for this book shows that a majority (88%) of leaders and senior managers responsible for cross-cultural teams believe that national culture influences knowledge-sharing practices for global innovation projects (Jensen 2014). The remaining study participants (12%) believed their organizational culture had succeeded in developing an environment where cultural differences did not influence knowledge-sharing practices.

Knowledge-Sharing and Multicultural Innovation

In exploring cultural differences, there are five elements identified within the knowledge-sharing process: **structure, openness, power, initiative, and response** (Jensen 2014). Structure relates to differences in how knowledge is organized and presented while openness relates to the degree and context of knowledge-sharing conducted in relation to the team environment. The perception and value of knowledge as power is closely linked to the ability to share knowledge within cross-cultural teams. Initiative relates to the comfort level for communicating new knowledge or ideas while response demonstrates the type of feedback that is required to ensure understanding, recognition and direction (Fig. 4.1).

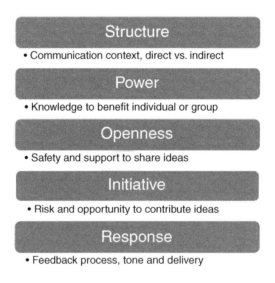

Fig. 4.1 Knowledge-sharing and cultural indicators

If we take a closer look at *structure*, there are differences in how knowledge is organized and presented during the project process where some cultures are considered more structured and formal while other cultures are viewed as less structured. North Americans and Northern Europeans were often described as more methodical and process-driven during project work, with a more linear and direct communication structure. However, the degree of formality in communicating between different levels of the hierarchy can vary, where there tends to be more informal communication in the US and variations of formal and informal communication across Europe. On the other hand, Asian team members were described as having a more holistic work and execution focus where a formal communication structure is more circular and indirect.

There is the notion of how knowledge is structured for communication where there is a more direct way to communicate as experienced in North America and Europe compared to a more indirect way to communicate as experienced in Asia. There is a Western tendency to present project information and knowledge as data and key learning points while in Asia you may see an emphasis on communicating one's experience with a particular topic. In Latin America, there tends to be a greater focus on story-telling and sharing client cases. A direct communication style is often framed in such a way where specific responses are sought such as "yes" and "no", while an indirect communication style may deliver a vague response such as "maybe" or "that depends". This has been the source of confusion for discussions around new concepts as well as project commitments within multicultural teams where "yes" can mean "maybe" or even "no" or "let's wait".

There is also the consideration of the organizational structure in how local market knowledge is shared globally within the organization. There can be a disconnect between the management team in HQ that keeps the knowledge at the global level and shares very little information with teams worldwide. On the other hand, local teams may keep knowledge in their own markets and do not share at the global level. Project and meeting processes therefore need to consider the global objectives and agenda while allowing more communication and exchange for local and regional team updates.

- A senior manager responsible for global product management at a multinational high-tech firm based in Asia shared her perspectives in managing planning efforts with teams across North America, EMEA, and Asia-Pacific, "I find (team members in) North America aggressive, they often exchange ideas frankly, though sometimes they seem a bit pushy, but they are really highly efficient and willing to take more responsibility in the cooperation process". She then reflected on other regions, adding, "Europeans are often conservative and slower in decision-making within their own group and the global team." Then she shared a few final thoughts on teams within the Asia region where she was based: "Chinese leaders regard relationship as a very important thing, sometimes too much consideration about people relationship where they may ignore problems or lose the chance to make even better achievement in business target. Asia Pacific teammates are more sophisticated and changeable where I need tactful communication methods…But they have a good team work spirit."

The perception and value of knowledge as *power* is closely linked to the degree of knowledge-sharing that occurs between teams. Cultures that display a strong hierarchical structure have difficulty sharing knowledge since it is viewed as power and ownership of expertise for the senior leader or manager that holds the knowledge. There is the sense of enhanced job status and job security if knowledge is held only by the individual. There is also the view that leaders should hold the most valuable knowledge for the project. This can influence knowledge-sharing in two ways: (1) The idea hierarchy which determines if the leader is the only initiator and provider of ideas or whether teams can participate in their development and (2) The decision hierarchy that determines the role of the leader in approving and supporting ideas proposed by team members. The leader has great influence in shaping power dynamics in keeping control of ideas and knowledge or empowering the team to share ideas and knowledge.

In most Asian cultures there is less encouragement of knowledge-sharing due to the view of knowledge as power in the organizational hierarchy. There are more expectations for leaders to have the authority for idea creation, knowledge-sharing, and decision-making which results in an innovation hierarchy of top down directives. On the other hand, North Americans tend to view knowledge as empowerment when shared with team members, especially if this action is recognized as a benefit to the innovation initiative. Knowledge is considered more valuable when it is shared and developed within the team. Team collaboration and sharing is encouraged in order to generate new ideas and information that lead to improved solutions. This is a view that is mainly supported in Europe yet there are variations depending on the countries. In general, there is an emphasis on team co-creation and knowledge-sharing, especially in Northern Europe. However, one can also encounter stronger idea and decision-making hierarchies in countries such as

France and Germany. It's important to recognize perceptions where knowledge can be viewed as power for the individual or knowledge can be viewed as empowerment of the team.

- A senior European product manager was more at ease in working with his Indian colleagues than his French colleagues due to the role of power and cultural differences in knowledge-sharing. Although the corporate view of knowledge-sharing promoted an open approach to all team members, the national culture could conflict in terms of local views of hierarchy, knowledge, and power. "This was more visible when I was working with our French companies where they have their decision process and everything needs to go up the structure for decisions. They took their decisions down to lower/micro levels…they kept knowledge and didn't share in order to benefit when working together so they knew what's going on locally. There were several times they didn't share customer knowledge in the area that I was responsible for, I found this very upsetting." While struggling to find a better communication system for sharing local customer and market knowledge in Europe, the senior manager found it easier to communicate during his visits to India. "The Indians, with all of them, it was very easy to communicate. I didn't have the behavior of not sharing information, they were very interested in discussions and sharing feedback." It's interesting to note that French culture tends to focus on knowledge ownership by leaders and key team members. On the other hand, Indian teams may not always indicate who owns the knowledge so there is a greater need for directions and details since the teams are focused on effective execution.
- A senior global product manager working for a European multinational firm found that he needed to constantly adapt to different knowledge-sharing practices of his multicultural and geographically distributed teams in Europe, Asia, and North America. He viewed his leadership role as a facilitator in improving the ability for the regional teams to bring new innovations. Speaking about the importance of leveraging strengths of different team members, he emphasized "for cross-cultural innovation, leaders need to be very aware of communication style and decision processes." When comparing cultures, he noted "for example, in France there are formal responsibilities in teams where you find knowledge which is owned by the person. They may not want to share since its their responsibility and

they feel they may lose their position." When considering other regions and countries, he added "In China, they listen carefully to the boss, and only share based upon the manager's directions…In North America, sharing is done when the immediate benefit is clear."

Power can influence the degree of *openness* to knowledge-sharing where the hierarchy can dictate the amount of information and the type of knowledge shared amongst team members. When there is more authority and hierarchy attached to knowledge-sharing, there is a tendency to share less and to provide knowledge linked to the project task without expanding on the context and related information. Openness requires a safe and supportive environment for encouraging open communication, knowledge, and new ideas from your colleagues around the world. The process for sharing ideas and knowledge needs to support a positive and collaborative view in building upon the ideas rather than using critique or judgment. There is the role of "saving face" in Asian cultures where a different idea or topic is not shared in order to avoid embarrassing someone or causing conflict due to opposing views.

There may be a tendency of conflict avoidance for many Asian cultures (including Japan, South Korea, and China) where the idea generation process may inhibit Asian members from providing ideas that could challenge other ideas due to fear of conflict. The American approach to spontaneous brainstorming in a very vocal and direct manner encourages debate of different ideas which does not always meet with the same enthusiasm in other regions, where Europeans may require more discussion and Asians may find it too confrontational. In referring to team frictions, a senior product manager explained: "You need to have a basic understanding of how colleagues work around the world and accept cultural differences. It may be viewed by others that Americans are 'pushy Cowboys' but we need to break that stereotype." Integrated idea generation methods that optimize diverse cultural perspectives work best for global teams.

For example, Americans and Europeans tend to practice an open and group-oriented approach to brainstorming and sharing knowledge where everyone is expected to contribute spontaneously and directly within an immediate time period. This places team members from many Asian countries at a disadvantage due to time needed to reflect on ideas as well as formulate them in the English language. Due to the different communication and language styles, team members from Asia and other regions with non-native English language speakers need more time to share and provide knowledge for the team project. Most team members from Asia appreciate time to think and reflect about the process in writing since they are uncomfortable in speaking before they fully understand the objective and the process. Deep reflection and focus is a practice that can greatly benefit cross-cultural teams from every region.

There is also the openness context of knowledge-sharing within multicultural teams. The American approach tends to focus on immediate sharing of team knowledge to ensure visibility and exchange whereas the Asian approach tends to be more conservative with knowledge-sharing which may be done indirectly and require more time. The European approach tends to find a middle ground depending on the cultures involved; there is both knowledge-sharing that is direct and open as well as more indirect and less open depending on the project phase. This is often linked to role and responsibility where knowledge-sharing in Asia and to some extent in Europe is linked to the role of the leader and team members. There is also the context of sharing in informal and formal settings, where certain cultures such as in Asia may share more openly in a social setting or team happy hour. In addition, the degree of transparency and openness to knowledge-sharing is influenced by the way knowledge is created, stored, and diffused by the project leader and the team in the organization. Both global and local knowledge should be shared and accessible to teams worldwide.

- A senior director and group program manager at a multinational firm with HQ based in Europe spent much of his time exploring, identifying, and responding to innovation market opportunities with cross-functional and cross-cultural teams. He found there was much competition between teams yet differences in how they shared information. "In the US it's normal to share at the early stages and promise what might be difficult to deliver. However, in Finnish and German cultures they may be more conservative in how and what they are sharing. Since they don't want to overpromise in delivery, the results should speak." Reflecting on his experience in Asia, he added "The Chinese may be more focused on local needs, information may not be so relevant due to specific local needs versus HQ needs."

Initiative demonstrates the ability to proactively share and contribute ideas as well as knowledge during project collaboration. It is also influenced by security in job roles and the work environment for team members. There may be fear concerning the ability of the initiative to succeed or fail and the resultant impact upon their job role and responsibility. In addition, there is the risk that a new initiative may take away resources that are dedicated to existing projects. Thus, local team members may show some resistance due to the overall stability for job roles and responsibilities.

There's also the entrepreneurial behavior of initiating new ideas that needs to be supported and promoted for team members to feel comfortable in following such actions. In Asian cultures, such as Japan and South Korea, taking initiative is not fully accepted nor practiced due to group consensus and managerial hierarchy. With internal knowledge platforms and social media tools, there has been a drive inside organizations to push employees towards "self-service knowledge". However, it's difficult to translate this behavior to other cultures since it requires adaptation of language and communication structures. In addition, other cultures may consider it disrespectful to share knowledge spontaneously and may not share knowledge unless requested. Leaders need to consider how to promote and support a safe environment for embracing innovation and new initiatives.

– A senior director working for a multinational high-tech firm in the US was responsible for a global team in launching new concepts to key international markets. Reflecting on cultural differences in knowledge-sharing, he exclaimed "Americans have a reputation of being in your face. For Europe, France and UK may also be direct. So when you're talking to someone in more reserved cultures, when you go to Asian cultures such as Japan and China, culture is a factor." He found that relying on web communication or conference calls was not sufficient for team meetings. "Whenever we have conference calls for tough decisions, western cultures such as US and Europe would speak up and you would think our Asian counterparts were in agreement because they didn't speak up. They then wait until the meeting is over and talk to each other, and then you find out later that they are not so happy with and not in agreement with the decision." In order to improve collaboration for the team, he suggested increased cultural understanding: "How you understand cultures and meeting rules, such as telling Western cultures to tone it down, and to allow Asian cultures to express more or share in such a way to provide important information."

When encouraging knowledge-sharing for project collaboration, *response* serves an important role in shaping how leaders provide and receive feedback. This is important for each of the innovation project phases, especially during ideation, strategic planning, and validation. Providing limited or insufficient feedback or framing it in a way that is perceived as too negative or positive can have an impact on local team engagement. When providing feedback, there is the tone and delivery as well as the type of information conveyed for ensuring understanding. The tone of the message and the way it is framed can influence cultural perception and motivation. The delivery is also important for ensuring a clear understanding of project objectives and strategy as well as alignment with teams worldwide.

For example, Americans often practice a "sandwich" style feedback process where there is always a positive response and praise followed by the critique and then concluded by a positive comment and encouragement. This may contradict the style that is often used by the French where immediate criticism is provided in feedback and then concluded by a more positive statement for future improvement and actions – this is seen as a more constructive critique and process. While they are both explicit and direct in tone, there may be confusion concerning the delivery as the American will find the French style too direct and negative with limited encouragement. On the other hand, the French may find the American feedback style superficial and too positive, lacking concrete actions for improvement. If a Chinese team member provides or responds to feedback, the tone will be implicit and indirect while the delivery is focused on providing examples and sharing knowledge directly related to the project. The feedback styles provided by the French and American colleagues would be considered too direct and aggressive by the Chinese team member, without sufficient information for project execution. On the other hand, the French and American colleagues will have difficulties understanding the nature of the feedback from their Chinese colleague since they are accustomed to more explicit and direct communication.

Special consideration to the nature of the response for new ideas and initiatives is important in building trust and motivation with multicultural teams. Attention also needs to be paid to the delivery of your feedback in terms of the type and amount of information that is shared. As in the above example, the North American and European cultures tend to focus on explicit and direct feedback with varying degrees of contextual information (from the US style of providing bullet point essentials to the European style of providing more examples). On the other hand, in Latin America, there is generally more emphasis on story-telling and providing vivid examples in expressive ways.

Consideration also needs to be given to the relationship of the feedback to the innovation project and the work process. Whether you share through examples, stories, theories, or facts, the delivery needs to be checked for cultural alignment. For example, a product management director who worked often with teams in India and Asia found that it was important to provide the big picture of the project while focusing on explicit directions and details to ensure successful execution. Roles and tasks need to be defined in order to make sense of feedback for the project process. Both tone and delivery should be evaluated for the appropriate communication vehicles, whether delivered in person, email, video conference, or social media.

- A global program director responsible for the launch of products and solutions at a high-tech multinational firm in Europe had enjoyed a long international career in working with over 20 countries. Reflecting on his experience and career, he noted the importance of gaining knowledge about the diverse regions and nationalities for effectively facilitating cross-cultural team collaboration. In providing examples, he found that "Japanese colleagues are not so open to talk about negative things, they want to embrace the positive. Yet, in Russia, people talk about problems more often and openly. If you ask what they don't like, they will have a list of what needs to be improved. However, with Americans they see positive things and lessons on what needs to be improved." In summarizing his experience, the program director noted that communication styles always need to be considered for global projects. "If you have people from different cultural backgrounds working together and not knowing about other cultures, it may result in conflicts. Each cultural background has a particular way in how to make decisions and how to collaborate."

Learning from Eastern and Western Perspectives

There are interesting contrasts in knowledge-sharing perspectives and practices between East and West. Most of the leaders interviewed for the global and local studies indicated that cultural differences in Asia in comparison to Europe and North America could be challenging when communicating and collaborating on innovation projects (Jensen 2014, 2015). As shown in

Table 4.1 Western and eastern cultural perspectives in knowledge-sharing for innovation

Western cultures	Eastern cultures
• Open structure and some formality	• Closed structure and formality
• Authoritative and shared leadership	• Leadership authority
• Direct, vocal expression	• Indirect, less vocal expression
• Flatter idea hierarchy, team focus	• Ideas hierarchy, needs to flow to top
• Sharing openly, in moderation	• Sharing less, in moderation
• "Thinking out loud"	• "Thinking and reflecting"
• Sharing spontaneously	• Sharing if requested
• Direct communication and sharing	• Indirect communication and sharing
• Challenge different ideas	• Avoid conflict/disagreement of ideas
• Focus on individual initiative and ideas	• Focus on team initiative and ideas

Table 4.1, the front end innovation phases of creation, planning, and validation require more consideration and understanding for cultural perspectives in creating, evaluating, and developing new concepts. Structure can play a role in terms of the communication context and the level of formality, a more open structure with less formality in the West compared to a more closed structure and formality in the East.

There is the question of power and the idea hierarchy, from the power of the leader and knowledge expert to shared knowledge and the empowered team. The degree of openness to sharing new ideas is influenced by both direct and indirect communication, from the spontaneous "thinking out loud" approach popularized by Western brainstorming to the more reflective and holistic approach in the East. A senior global product manager, with much experience in the Asia Pacific region, emphasized that "Asians are more thoughtful before they speak. It doesn't mean they don't have opinions or a point of view, they think more over the problem before verbalizing. When comparing this perspective with the US culture, he added "American culture promotes speaking before you think and this will evolve into open thinking and discussion. Some elements need to balance where you think through a problem and then work solution."

The perception of initiative-taking can influence the level of team engagement, where taking risk and creating opportunities can encourage more initiatives. However, avoidance of risk, conflict and fear of failure can discourage initiative-taking. This can influence practices that range from spontaneous sharing to sharing by request. The response of leaders to local team members should consider the feedback process for contributing and sharing knowledge, where cultural behaviors can vary from challenging ideas to avoiding conflict or "saving face" which is common in Asia. A senior manager responsible for international operations at a multinational firm with HQ in Asia explained that "for example, in China or Korea, people may not proactively share information or start a dialogue or conversation if it's not necessary. They avoid conflict or disagreement yet they share information when someone needs my support. If it's not needed, then they don't proactively share or start a discussion. The perspective is that you should not share if it's not verified."

Tone and delivery require more attention in terms of framing communication and knowledge-sharing while encouraging dialogue and global project collaboration. A global product line manager based in Europe reflected on his experiences in working with Americans, Chinese, and Indians to request feedback for key project topics. He observed that "Americans are more pragmatic, they will underline the bottom line and how to earn money. Chinese people will say yes even if they don't know how or know the answer while for Indians it's always possible to do it even if they don't know the topic." Another colleague working on a different product line added "My experience with teams in India and China shows that 'yes' and 'no' can have different meanings, you need to know how to translate."

While there may be differences in creating, initiating, and sharing new ideas and knowledge, there are also great opportunities to learn from diverse cultural perspectives and improve the global innovation process. The Asian philosophy and practice of allowing time for thinking and reflection when generating ideas brings more insights into the creation process. On the other hand, the American approach to open and creative thinking helps generate diverse approaches that can be integrated into the discussion. While the European style of blending creativity with reflection and discussion helps enrich the dialogue.

Every team dynamic is different and offers the opportunity to learn from the rich perspectives of personal, professional, and cultural experiences. Therein lies the opportunity to accelerate creativity and innovation from around the world. What is most important is to avoid reliance on an innovation process that may be ethnocentric or country-centric in its approach. Leaders need to introduce models and processes that enable the integration of different perspectives and practices. The key is to capture the diversity of ideas and approaches into a process that can optimize the talents and knowledge within multicultural teams.

> **Innovating Through Inclusive Dialogue and Culturally Diverse Teams at Intel**
>
> Regarded as a pioneer and leader in the Silicon Valley technology industry, Intel has evolved from computer processors to the cloud, the Internet of Things, memory and programmable solutions and 5G connectivity. This multinational firm has a worldwide presence with 107,300+ employees and offices located in 46 countries, including manufacturing sites, marketing and sales offices. Intel has its vision set on the innovation experience across geographies in optimizing multicultural talent through diversity, inclusion, education and sustainability.
>
> Intel's corporate values and organizational culture are firmly based on the strength of culturally diverse talent that collaborate across the global network to design and deliver innovative solutions. Through a newly launched program called GROW, the company intends to promote a growth mindset and inclusion for employees worldwide. The program relies on a set of shared resources and tools that impact behaviors for driving collaboration and innovation. Opportunities are always created for sharing cross-cultural knowledge. Within the organization, there have been programs that encourage cross-cultural learning such as "A Day in the Life" where employees from diverse countries provide training classes with speakers and immersive learning about their culture. This allows everyone to share and celebrate their cultures and customs while enjoying international cuisine.

Attention to multicultural learning and collaboration is an essential part of Intel's innovation projects. With global teams, market segments, and product lines scaling international markets, there is the need to show local responsiveness for the global strategy. There is always the challenge of project execution and scaling processes which makes it important to explore different approaches to optimizing project collaboration. Aiming for global integration and local responsiveness, there is the opportunity for team collaboration on cultural, functional, and social levels.

In creating a more cohesive and united global team, close collaboration and functional integration are encouraged when developing and delivering new concepts and solutions. Paired with their counterparts in other countries, project teams can collaborate on designs, implementation, and reviews. Both global and local team members have ownership and accountability while they are allowed to resolve problems at the local level to effectively manage time and communication. This process allows them to move from one phase to the next phase for improved timeliness.

Intel has a long history in developing leadership programs with a cultural focus that include training in cross-cultural management, country knowledge, and foreign languages. Cultural integration within project collaboration demonstrates common Intel values through cross-cultural awareness seminars and round-table discussions that help reduce cultural misunderstanding and unnecessary conflict. Social integration is achieved through cross-site visits between HQ teams and local teams in order to share the culture and understand the local environment. Local team members introduce their culture, get to know their counterparts socially, develop trust and positive cultural views. Most importantly, there is the opportunity for face-to-face project collaboration as well as social gatherings outside of work which strengthen relationships within the global team.

Benefits from multicultural team collaboration include the importance of face-to-face project collaboration and cross-site visits for developing cultural understanding and long-lasting work relationships. Organizational development has been improved through global learning and skills development, local customer knowledge and sharing, leadership and influencing opportunities at global and local levels. Cultural and social integration is just as important as functional integration. Equality amongst team members across geographies as well as reciprocity are key factors in successful team collaboration, in addition to an effective and open communication process. Intel's focus on optimizing global innovation through multicultural collaboration has created a competitive advantage and international market growth for its product portfolio.

Sources: Company Overview, Intel: http://www.intel.com/content/www/us/en/company-overview/company-overview.html. Accessed on December 2, 2016.

Danielle, Brown, Chief Diversity and Inclusion Officer, Message, Intel Works to GROW Diversity and Inclusion, Intel: http://www.intel.com/content/www/us/en/diversity/message-from-danielle-brown.html. Accessed on December 2, 2016.

Frauenheim Ed. Workforce Magazine, Crossing Cultures, December 2, 2005: http://www.workforce.com/2005/12/02/crossing-cultures/.

Intel Annual Report 2015. Intel Investor Relations: https://www.intc.com/investor-relations/financials-and-filings/annual-reports-andproxy/default.aspx. Accessed on December 1, 2016.

Jensen, Karina R. Global Innovation and Collaboration Study, 2014.

Life at Intel page, Intel: www.intel.com/content/www/us/en/jobs/life-at-intel.html. Accessed on December 2, 2016.

Weisul, Kimberly. "This is What Happens When A Big Tech Company Gets Serious About Diversity", Inc. April 19, 2016: http://www.inc.com/kimberlyweisul/what-happens-when-a-big-tech-company-serious-diversity.html.

Creating Global Dialogue Through Local Knowledge-Sharing

- Introduce a meeting agenda where each team member can take the time to prepare and share an update, a success story, or new market insight from their countries.
- Alternate time zones for international team meetings with multiple geographies, i.e. allow the team to accommodate their world time clock with video conferences held in Asia-Pacific to Europe, Middle East, Africa to North America and Latin America.
- Organize team learning sessions through cultural integration and project collaboration, where team members from HQ and local subsidiaries team up and contribute to cross-cultural awareness seminars and discussions for improving cultural understanding, from creation to execution.
- Start with a "silent brainstorming" or reflection exercise prior to the ideation session where you allow each team member time to think about new ideas and contributions – this helps create a dialogue where everyone can contribute their ideas.
- Use a knowledge-sharing process that enables the integration of different cultural perspectives during discovery, ideation, concept creation, and planning activities.

Bibliography

Adler, Nancy J. *International Dimensions of Organizational Behavior*. Cincinnati: Southwestern College Publishing, 1997, 2007.

Cohen, W.M. and Levinthal, D.A. "Absorptive Capacity: A New Perspective on Learning and Innovation", *Administrative Science Quarterly*, 35(1990): 128–52.

Foss et al. "Governing Knowledge Sharing in Organizations: Levels of Analysis, Governance Mechanisms, and Research Directions", *Journal of Management Studies*, 47:3 (2010): 455–82.

Hall, Edward T. *Beyond Culture*. Garden City, NY: Anchor Press, 1976.

Holden, Nigel J. *Cross-cultural Management: A Knowledge Management Perspective*. Great Britain: Prentice Hall, 2002.

Jensen, Karina R. Global Innovation and Collaboration Study, 2014.

Jensen, Karina R. Local Innovation and Collaboration Study, Asia Region, 2015.

Nonaka, Ikujiro. "A Dynamic Theory of Organizational Knowledge Creation", *Organization Science*, 5(1994): 14–35.

Piekkari, Rebecca et al. "The Multifaceted Role of Language in International Business: Unpacking the Forms, Functions, and Features of a Critical Challenge to MNC Theory and Performance", *Journal of International Business Studies*, 45:5 (June 2014): 495–507.

Tsai, W. "Knowledge Transfer in Intra Organizational Networks", *Academy of Management Journal*, 44(2001): 996–1004.

5

Listening to Local Market Voices

"Our organization is moving at a fast and exciting pace, but we're afraid of missing out on opportunities if international teams don't have sufficient exchange and knowledge." A director responsible for global strategic programs at a multinational information technology firm with HQ based in Asia was very excited about his new mission. He would be responsible for integrating the product innovation process, from new concept development to international market introduction. The position involved a global scope and close collaboration with cross-functional teams in HQ as well as cross-cultural teams in key geographic locations. However, the new assignment soon faced great challenges as the leadership teams and organizational culture were not aligned with local market innovation and collaboration. Although the firm had gained a reputation for innovation worldwide, it relied heavily on centralized research and design at HQ. There was also a lack of cultural and market understanding by the HQ executive team to listen, respond, and adapt to new ideas proposed by local teams around the world.

"There have been incidents where the HQ leadership team develops the global plan and shares it with subsidiary teams," the director explained, "the local teams are then frustrated with the plan and communicate that it doesn't work for their markets." The situation became even more difficult when the project leads did not communicate feedback and requests from local teams to the HQ executive team – there was a lack of interest to listen and they did not want to disrupt the global plan. In addition, there was the role of national culture at HQ in having fear of questioning the leadership team since they had issued the plan. The new

© The Author(s) 2017
K.R. Jensen, *Leading Global Innovation*,
DOI 10.1007/978-3-319-53505-0_5

concept was launched globally and product issues occurred resulting in poor market results at local levels. "The project team did not want to share local feedback with the leadership team in HQ since they wanted to get the project done per executive orders," the director noted with frustration as he added, "the local markets were dealing with local knowledge and customer needs while HQ was dealing with its own determination to get the product to market."

In seeking a more collaborative process moving forward, the director noted that "we need to have the HQ leadership teams more open to listening." With improved local market understanding and response, there was an opportunity to increase local market success and increased respect throughout the organization. A new initiative focused on relationship-building across functions and cultures. Additional programs focused on country assignment rotations for becoming local and regional experts in order to qualify new ideas and to better understand local teams and markets. In creating more cultural awareness across the organization, the director hoped to increase alignment and collaboration with teams worldwide.

Facilitating Multicultural Team Collaboration, from Concept to Market

Multicultural team collaboration is not an easy endeavor and may be best suited for those who enjoy a global challenge involving extreme networking, multi-tasking, and travel adventures. When leading cross-cultural teams, leaders and managers are most challenged by trust-building, team participation, interactive dialogue, team understanding and support of the global strategy (see Fig. 5.1) (Jensen 2011). Trust-building, shared goals and team

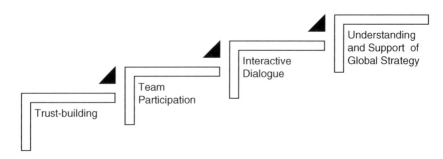

Fig. 5.1 Challenges in multicultural team collaboration

creativity are a priority during the pre-launch or planning phase while knowledge-sharing and conflict management need to be managed throughout the product launch process. When managing cross-cultural teams, it is especially important to establish trust and encourage contribution of ideas when the project is in the planning phase. Strong trust and open communication amongst team members facilitate conflict management and knowledge-sharing throughout the planning and execution phases.

When leading a global innovation project, the senior manager is faced with the objective of creating a strategic plan that effectively differentiates and positions the new concept to key geographic markets. The ability to understand the value and relevance of the new concept for countries and regions relies on the local market knowledge held within the organization. Moreover, the local team members often possess the necessary market and customer knowledge that is required for identifying and validating local solutions. With an increased focus on communication technologies and reduced focus on travel, leaders are confronted with the challenges of building trust and developing relationships across geographic and cultural distances, while trying to ensure a balance of face-to-face interactions and virtual communication on a limited budget.

Global innovation projects demand more time for social interaction and connection within the team, including in person meetings and visits to HQ and subsidiaries, special team-building events, and project presentations. In facilitating multicultural collaboration, there is the necessity to carefully manage commitments for international product and service innovations in delivering on promises made to local teams and customers. It requires team leaders to provide a consistent experience in acting on their promises, showing actions, and delivering on their commitments. When there's consistency between words and actions, there is increased credibility which leads to trust-building across cultures and functions. Throughout the global innovation project cycle, leaders need to ensure meaningful work and well-defined tasks for geographically distributed teams.

As emphasized in Chapter 2, the greatest challenges for leaders in facilitating knowledge-sharing and contribution from cross-cultural teams involve global market visibility, the strategic planning process, local market intelligence, team communication and knowledge-sharing (Jensen 2014). The roles of HQ and subsidiaries in strategic planning are often hampered by the organizational structure and culture, time pressure, the project collaboration process, and the assumption that strategy is HQ driven while subsidiaries are execution-focused. This is further complicated by ineffective local market intelligence skills with lack of cultural understanding and limited access to local teams and markets.

On the other hand, the issue of global visibility applies to local teams and subsidiaries where they lack a complete view of HQ and the global network as well as an inability to present the local business case. This is further challenged by the lack of an effective feedback system that captures and responds to local needs. There is a greater need for communication that considers cultural and language differences through a dedicated knowledge-sharing system. The lack of effective planning and communication processes support the need for more knowledge-sharing throughout the global innovation cycle, from planning to execution.

Serving as liaisons and facilitators for the conception and execution of new concepts to international markets, global project leaders have unique insights to perspectives of HQ and subsidiary management teams around the world. In order to validate their views, let's take a look at local views and perceptions from regional teams in Asia. A summary of perceived challenges by global and local team leaders is provided in Table 5.1. The greatest challenges to facilitating knowledge-sharing and contribution from the local view include *communication and knowledge-sharing, project process, local involvement, local market understanding,* and *lack of resources* (Jensen 2015). In comparing these views with those of global project leaders, there's agreement on communication and knowledge-sharing, project process, local involvement, and local market understanding. What is particular to local needs is the availability of sufficient resources and time for developing new market opportunities.

When addressing the challenges of communication and knowledge-sharing between HQ management teams and local subsidiaries in Asia, there is an emphasis on the lack of communication, the context for knowledge-sharing, the lack of opportunity for risk-taking and knowledg-sharing, the lack of a clear structure and process, and the need for more cross-cultural team interaction at the same site. Improved communication through knowledge platforms and interpersonal dialogue between teams in HQ and subsidiaries are a priority. The context for knowledge-sharing requires more cultural awareness and understanding of local market needs and requirements.

In order to encourage more knowledge-sharing, the project leader needs to consider how to allow time and space as well as encourage behaviors for risk-taking and knowledge-sharing. The lack of clear roles and a communication process can have a negative impact on the global team's ability to build trust and relationships where knowledge can be shared and captured. On the other hand, local teams may need to make more effort to initiate ideas and raise issues that can impact concept creation and market

Table 5.1 Perceived challenges by global and local team leaders

Global challenges with subsidiaries *Global project and team leaders/HQ*	Local challenges with HQ *Regional and local team management/Asia*
• Communication and knowledge-sharing • HQ and subsidiary roles • Local and global visibility • Local market intelligence • Effective feedback system	• Communication and knowledge-sharing • Project process • Local involvement • Local market understanding • Lack of resources
Views of global project leaders/HQ	**Views of regional and local team leaders/Asia**
• "HQ doesn't involve subsidiaries in early stage" • "Executives are interested in numbers, not process" • "Executive management doesn't understand local market needs" • "There is less understanding when you move up the executive chain" • "Lack of resources to satisfy regional needs" • "We need to give regions, countries, and customers a chance to voice their needs and interests" • "The innovation process needs to share knowledge about international markets" • "Feedback has a disconnect between HQ and subsidiary priorities" • "We need to establish a culture of trust with local markets and find a common language" • "More time is needed to gather knowledge"	• "HQ may not understand local needs or initiatives that deviate from global model" • "People in HQ act like they know more" • "HQ needs to involve the local team" • "Local market voices are not heard so we're not eager to contribute" • "Send more local teams to HQ and HQ teams to local subsidiaries" • "Having face-to-face communications and ensuring exchange" • "Stay open-minded and be more transparent" • "We don't get attention from management, larger markets dominate" • "More ownership and closeness to market" • "It takes a lot of effort to sell and persuade with new ideas, it's a very big process"

opportunities. Co-locating teams and encouraging more visits and exchange with local team members can improve project collaboration in international markets.

In summary, both global and local team leaders feel that current communication and knowledge-sharing practices could be more structured and organized while allowing for more time, space, and transparency for enhancing multicultural team collaboration. This is especially due to the tensions caused by the need to pursue market exploration while sustaining market exploitation for existing products. There is also the ongoing tension between team members in HQ and local subsidiaries concerning their roles and level of involvement in the global innovation process. Increased understanding and involvement of HQ senior management is necessary for recognizing and rewarding local market opportunities.

Local management team views in Asia that differ from those of global project leaders are focused on the lack of local involvement in the planning process and insufficient resources. Local team managers find that HQ management teams don't have sufficient understanding nor make sufficient effort to understand local market and customer needs. There is a perception that HQ management teams are too focused on the global strategy without considering local market requirements. With a lack of a global view and understanding of the international strategy, there is clearly a need to improve dialogue in order to accelerate multicultural collaboration and innovation.

Building Trust Across Cultures and Functions

The opportunity for leaders and teams to learn from local market knowledge can contribute to improved solutions for international markets. Diverse cultural perspectives can contribute to a more vibrant dialogue and improved innovation performance. Teams can discover and learn about cultural and market knowledge that respond to international customer needs. Yet multicultural collaboration and knowledge-sharing require trust between culturally diverse and geographically distributed team members. Trust is found to be the most challenging element of leading global teams and ensuring successful project results. In exploring the key elements of trust-building for multicultural innovation and collaboration, five themes have been identified – **relationship-building, listening, response, open communication, and project contribution** (Jensen 2014).

Relationship-Building

There is the need to provide more time for social interaction and connection through *relationship-building*. This includes meeting with people in person through visits to HQ and subsidiaries, special team-building events, and strategic planning sessions held in local markets. Local team members need to be updated on recent news and practices where a senior manager emphasized the need to "share and show new product innovations and have discussions". Showing an openness to learn and understand local team members can build the cultural bridges needed for developing a global network. The ability to socialize through informal settings helps establish personal connections through mixers and celebrations.

Listening

The opportunity to capture perspectives of diverse cultures requires the ability to *listen* to local customer and market needs. Acquiring a local understanding and perspective demonstrates consideration for local and regional needs. By listening to local teams, you are allowing their voices to be heard and empowering them to provide value to the project. It also creates an environment that is inclusive in welcoming and recognizing cultural and market knowledge from team members. As expressed by an international product director "you need to understand their point of view and their markets".

Responsiveness

In order to show integrity as a global team leader, the ability to *respond* and show action on promises may be the most critical element. There is the importance of following through on a promise or statement made to local teams in preparing the global product launch. The project requires transparency and accountability through regular communication and delivery on promises. This is especially important when project leaders need to respond and provide information for proposals, requests, and plans for local markets. It's the ability of project leaders to provide a consistent experience to teams in acting on their promise, showing action, and delivering on their commitments. When there is consistency between words and actions, there is increased credibility which leads to a faster establishment of trust. As noted by a senior manager, "People trust you if you follow up on your word – it gains respect and commitment." Active communication and feedback between leaders and teams can contribute to increased trust.

Open Communication

Transparency and *open communication* go hand in hand when working on complex multicultural innovation projects. Open and honest dialogue should be nurtured through face-to-face interactions and meetings early in the project phase. Frequent communication is ensured through a combination of on site and online communications from meetings to web and video conferences. In order to ensure sufficient attention and commitment to regular communication, an effective agenda with strong incentives are recommended. Regular face-to-face meetings combined with online communication tools on a regular basis creates a strong platform for inclusive dialogue.

Project Contribution

The role of *project contribution* is critical to reinforcing the value of team members through collaborative innovation. Global project leaders need to ensure that team members have meaningful work and well-defined tasks during the global innovation project. Collaboration with the worldwide team provides local team members with the opportunity to enhance their recognition and contribution to strategic initiatives. It is the leader's role to provide teams with meaningful work while guiding the process and facilitating any challenges or conflicts. Team members trust the process when they are valued and recognized for their role and contributions on the innovation project. As noted by a global program director: "Build a win-win collaboration". Project collaboration from idea generation to global launch provides an opportunity for increased trust in recognizing the value and knowledge of teams around the world.

The Local View on Trust

The experience and perspectives of global project leaders show the need to build relationships, to listen and respond, to enable open communication and project contribution as necessary ingredients for creating trust. However, would this also align with the view from local teams? Research with local and regional team managers in Asia explored how trust could be improved for their teams during global project collaboration (Jensen 2015). The study showed clear emphasis on the importance of a *cross-cultural mindset* in building trust amongst local teams. This finding is most closely linked to relationship-building and the ability to listen in order to have a local understanding. A cross-cultural mindset and behavior is the ability to gain knowledge and understanding of other cultures when building trust.

What is most interesting is that both *open communication* and *response* are shared between global project leaders and local teams. Open communication is the ability to drive transparency and conduct open and honest dialogue with local team members. It's important to provide sufficient knowledge for local team members to understand the context of the global innovation project and how they can contribute to the front end innovation process. As noted by a senior manager based in Vietnam: "How you share and the process of knowledge-sharing is important. In Viet Nam, we like to share openly through discussions rather than documents." Cultural preferences in Asian countries should be considered in order to create a knowledge-sharing process that is open and transparent.

The ability to act and deliver on promises has significant impact on the trust and confidence placed in the team leader since it's linked to project collaboration. "If you've made the commitment, such as a new product that you're trying to roll into the market, and you don't deliver as promised, then you break down trust for the team and customers. They will be very reluctant to collaborate in the future," explained a senior product manager based in Singapore. In order to establish credibility and authority, you need to understand the perspective of local teams, be honest about expectations, and ensure transparency throughout the project process. It's the ability of the project leader to recognize, respond, and deliver on requests and initiatives that will facilitate collaboration with local teams.

There is also the need to understand the cultural context and the power structures that are particular to Asian countries. As noted by a senior executive responsible for the Asia-Pacific region: "The biggest challenge is the cultural context and power structures – the strong power distance. It's hard to be successful if the cultural context and personality is not understood by team members. You need to select people that have certain mindsets or behaviors that allow them to interact well with a cross-cultural team." The ability to show understanding for the culture helps create the mindset and behavior that builds trust when interacting with local teams in Asia and other regions around the world.

Motivation and Multicultural Team Collaboration

The opportunity to build trust is closely aligned with team motivation and participation in the global innovation project. Global project leaders based in HQ identified specific behaviors that motivate multicultural team collaboration during planning and execution phases. There are five main themes that influence motivation – **recognition, responsiveness, listening, engagement, and local connection** (Jensen 2014). Increased recognition, responsiveness, and engagement motivate local team members to improve knowledge-sharing and contribute to the global innovation project.

Recognition and Engagement

Recognition and *engagement* are the most critical factors for facilitating knowledge-sharing during the global innovation project cycle. There is a need to recognize local team members' knowledge, talent, and expertise for contributing to front end innovation. In addition, *engagement* provides a sense of ownership within the innovation process which needs to be established in the early phases of planning, ideation, and validation. This requires responsiveness as well as more transparency and feedback concerning initiatives and requests. In speaking to the importance of engagement of team members for project collaboration, a local team manager based in Japan gave an example of creative opportunities that enable the team: "Last summer, we decided to add more critical features (for the local product) which involved rapid change. There was not much time for execution. Creativity was high. We were very creative and free to tackle new approaches and ideas."

The *recognition* of local team members' knowledge and expertise in the global launch project as well as the value their participation brings to the success of the project are key drivers. As noted by a senior manager: "If they make the effort to give us feedback and share their knowledge with others, (they're motivated) that we actually incorporate their feedback into our plans, and that other regions would adopt some of the best practices they share." The project leader needs to acknowledge contributions from team members and show how it has been applied to product design and development. A regional executive responsible for Asia-Pacific also emphasized the need for HQ management teams to focus on local markets: "You need to make sure they're (local teams) recognized and they matter." The team members aspire to be a part of the global team where they are recognized for their contributions through increased visibility and collaboration.

Responsiveness

In a fast moving and turbulent marketplace, local teams expect a speedy response to their requests for information, training, and support. Being responsive and having an effective process that ensures knowledge-sharing and local market performance is a motivation booster. Providing an inclusive process for new concept planning and execution creates more confidence in international market success. Leaders should provide consistent feedback to global teams regarding their ideas, concepts, and initiatives. In an organizational environment that supports multiple markets, there may not always be the opportunity to respond to or endorse every request; however, leaders need to provide a clear feedback process that communicates the outcome from team contributions as well as their reasoning for not acting.

Listening

In a continuously connected and technology-driven business environment, *listening* has become a precious skill for building motivation and engagement in mission-critical projects. Mobile tools are an integral part of business communication yet have also created challenges for interpersonal communication across cultures. Active listening remains the most important behavior for creating both trust and motivation amongst cross-cultural teams. Leaders truly need to listen to local market voices and show that team members' thoughts and ideas are valued within the organization. This competency emphasizes the leader's role as knowledge facilitator in order to encourage more communication and knowledge-sharing.

Local Connection

The ability to analyze and understand regional market needs requires a *local connection* in terms of cultural and market understanding. Moving beyond domestic market needs, international customers and markets demand increased attention and research in order to understand particular practices and preferences. Leaders need to ensure that local teams are part of the planning phase in order to share critical data about customers, markets, and competition. Leading and managing a global innovation project means frequent travel between HQ and subsidiary locations in order to understand country environments. Meeting face-to-face to develop relationships and share experiences is invaluable for developing relevant ideas and solid business proposals for successful execution.

The Local View on Motivation

In comparing themes identified by global project leaders in HQ and local managers in Asian subsidiaries, recognition and engagement are aligned with both groups in their importance for increased knowledge-sharing (Jensen 2015). Recognition and engagement are the most critical factors for facilitating knowledge-sharing during the global product innovation process. It is about recognizing the local team member's knowledge, talent, and expertise for contributing to global product innovation. In addition, engagement provides a sense of ownership within the innovation process which needs to be established in the early phases of planning, ideation, and validation. Global project leaders emphasized responsiveness while local team managers in Asia emphasize open communication. Global project leaders refer to responsiveness and more transparency concerning initiatives and requests, whereas local team managers refer to transparency and more knowledge-sharing.

The opportunity to apply local ideas to global opportunities reinforces the value of local market knowledge as an organizational resource. Local team members simply would like to see the impact and results of their contributions and how they can add value to the team project on a global level. A senior product manager based in the Chinese subsidiary of a US MNC emphasized the importance of making a global impact through the ability "to develop something that becomes a global feature, to contribute to the global market beyond the local market. If we provide the opportunity to share and to achieve more and collaborate with the global team, then we're more motivated." The senior manager explained the true reward is the recognition of the work and having a greater impact with new ideas and products on a global level. A senior manager working for the Vietnamese subsidiary of a European MNC emphasized that "you need to encourage members to use tools for sharing knowledge in the company worldwide, somehow they are not familiar with tools and never join a forum or discussion – we need to know how".

What is most interesting about the findings on trust and motivation is the emphasis on listening and responsiveness. As shown in Table 5.2, the secret formula is to listen and respond. In building trust and increasing motivation for cross-cultural teams, leaders should pay attention to the art and practice of listening and responsiveness. Demonstrating cultural and local market understanding through active listening and learning shows consideration and respect for knowledge within the team.

Table 5.2 Trust and motivation behaviors for multicultural team innovation

Trust	Motivation	Trust & Motivation
• Relationship-building	• Recognition	• **Listen**
• Listening	• Listening	• **Respond**
• Responsiveness	• Responsiveness	
• Open Communication	• Local Connection	
• Project Contribution	• Engagement	

Country visits and international meetings show that leaders value team members and their voices are heard. As expressed by a project director based in India: "Early engagement is best, there's more ownership for the teams and then they can take over the project." A senior manager based in China emphasized that "it's about trust between the HQ team and the regional unit, we need to build up trust, regional organizations and capabilities. For the local team, it's about sharing know-how since they don't provide or share much." A director responsible for the Chinese market added that "local teams are motivated to contribute but de-motivated by HQ actions. The type of response or lack of response from HQ management de-motivates people."

Listening is only as powerful as your response to local teams. While listening is critical for cross-cultural dialogue and understanding, there is also an expectation to act on promises and commitments. With a 24/7 global marketplace, organizations and leaders need to respond quickly and provide the necessary support to capture market opportunities. This requires an inclusive process from planning to execution with built-in check points for knowledge-sharing and feedback.

Essilor's Cross-Cultural Lens for Viewing the World

With a rich tradition dating back to 1849, Essilor started as a workers' cooperative of eyeglass makers in Paris and today offers global solutions for correction, protection and prevention of visual health. It offers a line of quality products and services dedicated to good vision through innovative business models, including ophthalmic lenses, sunwear, reading glasses, equipment and instruments. With headquarters located in France, Essilor has over 61,000 employees across 63 countries with R&D centers, plants, distribution centers, and prescription laboratories in key world regions.

The company's mission is simple yet powerful: improving lives by improving sight. Its culture and values are based upon working together, innovation, respect and trust, entrepreneurial spirit, and diversity.

Innovation is central to Essilor's mission where it seeks to explore solutions for diverse visual health needs in mature and emerging markets. This includes lenses designed for a connected life, vision care services and connected devices, and accessible vision care for impoverished regions in emerging markets. A global market leader with products in more than 100 countries, Essilor has developed a partnership strategy that is focused on local responsiveness through attention to specific contexts of customer solutions in locations worldwide.

Essilor has succeeded in creating an efficient project management process that integrates regional needs with clear prioritization and focus on resources. However, the company also needs to ensure increased agility and responsiveness for local markets. There is a focus on continuous improvement for the global innovation process as well as talent development to ensure knowledgeable, proactive, and customer-oriented team members in all locations. Since the company values have always been based upon collaboration, it has managed to develop an international mindset through culturally diverse teams and empowerment of local management.

The company supports open communication in order to create more transparency for teams within the global network. There is an appreciation for knowledge-sharing and recognition of local team initiatives. Regular travel opportunities combined with collaborative technologies facilitate knowledge-sharing for global project leaders and international teams. With a global vision, Essilor is successfully facilitating multicultural collaboration throughout the organization.

Sources: Company overview, Essilor Group: https://www.essilor.com/en/the-group/. Accessed on November 11, 2016.

Jensen, Karina R. Global Innovation and Collaboration Study, 2014.

Collective Wisdom through Collaboration

Leaders, teams, and organizations need to consider a balance of face-to-face communication with knowledge platforms that can harness and diffuse the collective wisdom of talent worldwide. Cross-cultural teams should have the knowledge, training, and resources that allow them to effectively initiate, communicate, and propose new ideas. In building a business case for persuading senior management teams to consider a new idea or initiative, geographically distributed teams need a knowledge-sharing structure that facilitates communication throughout the global innovation cycle. In the Essilor case, the focus on its organizational values of collaboration, diversity, and innovation combined with empowerment of local teams resulted in an open communication and knowledge-sharing process.

Although there is an increased interest and need to use communication technologies and collaboration tools, a comprehensive and integrated knowledge-sharing platform needs to support in-person communications. The opportunity to have a dashboard with a global view of communication and collaboration tools appear to be at the top of the wish list. Team work spaces combined with video or conference calls can be effective solutions. Standard templates and tools that promote a virtual work space are helpful in promoting knowledge-sharing and contribution. Technologies that can act as mechanisms for facilitating regional and cultural dynamics of innovation through a blended learning environment may be the best platform for promoting multicultural collaboration.

The need for improved internal collaboration with local markets is most apparent in the automotive industry which is experiencing a wave of disruptive innovation that will shape the future of transportation, from smart technologies to autonomous cars. In order to remain competitive and relevant with customers, car makers face the pressure of integrating both radical and incremental innovations in their product range. The rapid evolution of communication technologies and intelligent systems pose a challenge for automotive manufacturers due to longer launch timelines (an average of 5–7 years) for automobiles. When it's time for the global launch, every aspect of a new model needs to reflect the latest innovations.

Effective planning and timing for new product introductions is a necessity in responding to customer expectations for both automotive performance and state of the art technologies that enhance the driving experience. In evaluating two key players, BMW and Mazda, it is apparent there is more attention to local market needs and the engagement of local teams worldwide. Applying different approaches, their efforts focus on increased collaboration and communication throughout the global innovation cycle,

The BMW Group is a leading premium manufacturer of automobiles and motorcycles as well as provider of premium financial and innovative mobility services. With headquarters based in Munich, Germany, it operates production and assembly facilities in 14 countries and has a global sales network in more than 140 countries. The company has enjoyed consistent growth and expansion in mature and emerging economies, with primary markets that include China, Germany, and the US. It has a strategic focus on efficient dynamics and driving pleasure for maximum performance and minimal fuel consumption.

Mazda's brand essence is about "celebrate driving" and its corporate vision emphasizes "the Doh" ("way" or "path") of creativity for embracing challenges and optimizing innovation. With headquarters in Hiroshima, Japan, it is a leading manufacturer of sports cars, passenger cars and commercial vehicles to key world regions. With over 44,000 employees worldwide, Mazda has research and development sites in Asia, Europe, and North America as well as 144 sales offices outside of Japan. Its monozukuri innovation strategies focus on aiming high above the industry standard with attention to model and preference variations in mature and emerging markets as well as developing countries.

For BMW, the main challenge appears to be a focus on quality and performance excellence while ensuring responsiveness and efficient execution into key growth markets such as China. Due to the high growth rate experienced in China and emerging markets in Asia, customers are making new requests and there is demand for adaptation to local market needs. This requires more interaction with the management team in HQ where local teams can add value through their knowledge and experience. Front-end innovation is driven by a global concept with local adaptation depending on input from local teams.

More attention is paid to project collaboration and strategic co-creation since local teams from key markets are involved in the planning and validation of the global innovation strategy through global forums held in HQ. In order to prepare for the upcoming launch, the execution phase involves exchange through country visits and workshops that allow teams to share knowledge and socialize. From enjoying a test drive of the new model to exploring new ideas, the teams can share the BMW experience.

For Mazda, there appears to be a similar challenge where there is a need for increased communication and knowledge-sharing between executives and management teams in HQ and local management teams and dealers in the regions. Management is renewing their commitment to knowledge-sharing and collaboration through a dedicated innovation office, with attention to local customer preferences. The company has established a principle through "tomo iko" in learning from each other and sharing knowledge. There is increased international mobility with opportunities for exchanges, rotations and visits between headquarters and subsidiaries in regions worldwide.

Although there is a proliferation of social media and new technology tools available within the enterprise, there is still a preference for face-to-face communication and knowledge-sharing as the most effective methods for building trust, understanding, and collaboration. This requires organizations to offer more travel and exchanges between HQ and subsidiaries in every geographic location. As shown in the Adobe case on empowering team collaboration, social media and technology tools serve a complementary role in sustaining the dialogue and building relationships. Tools that enable team communication across geographies include video conference or telepresence, a knowledge-sharing and collaboration platform, a shared drive or web conference system, a project or web portal, and wiki type tools.

As discussed in this chapter, trust and motivation are powerful factors for creating effective multicultural collaboration and successful project results. The ability to listen to local market voices requires attention to cross-cultural learning and relationship-building through increased international mobility. Organization structures and dynamics between HQ and subsidiaries require improved connections within the global network in order to strengthen open communication and transparency. The roles and engagement of team members in project collaboration during the global innovation cycle are critical for optimizing knowledge and talent worldwide. Leaders have the opportunity to improve intercultural dialogue through active listening and responsiveness to international teams and customers.

Connecting Adobe's Global Network and Empowering Team Collaboration

Adobe is leading the era of online collaboration with its vision of changing the world through digital experiences. It offers marketing and document solutions that deliver digital creations to empower the user experience. Adobe believes in creativity as a catalyst for positive change and emphasizes corporate responsibility through education, sustainability, and community. With headquarters based in California, USA, it has a global reach with presence in 27 countries and 14,000+ employees worldwide.

With the increasing importance of customer and market knowledge, the challenge is to facilitate local team collaboration while ensuring agility and responsiveness. There are time constraints and the focus on execution efficiency that may limit participation in the front end. A global reach can be challenging when communicating across geographies and time zones. Leaders need to ensure sufficient time and space for new ideas and knowledge from team members worldwide.

In developing a strong focus on idea generation and creativity, global team members have the opportunity to listen and learn from local teams and customers throughout the global product innovation process. While there is a focus on co-creation, there is also an emphasis on responsiveness combined with effective execution. A collaborative environment creates an open and shared dialogue that empowers the global team to initiate and contribute ideas.

In order to facilitate multicultural team collaboration, Adobe is optimizing its own tools and collaborative technologies such as AdobeConnect and an internal portal with video content that can be shared across regions. Global teams can access industry expertise by sharing data and information that brings additional insights on how to optimize experiences from other cultures and markets. There are opportunities for building relationships and collaboration through country trips and global meetings that enable knowledge-sharing and feedback. Through real-time dialogue, these exchanges allow local teams to rapidly adapt and make changes for successful market results.

Sources: About Adobe page, Adobe: http://www.adobe.com/about-adobe.html. Accessed on November 16, 2016.

Company profile, Adobe: https://www.adobe.com/aboutadobe/careeropp/about.html. Accessed on November 16, 2016.

Jensen, Karina R. Global Innovation and Collaboration Study, 2014.

Global and Local Team Development through Cross-cultural Learning

- Team members can immerse themselves in the local culture through 3–6 month rotations where opportunities are created for sharing cross-cultural knowledge.
- Exchanges or extended country visits to enable learning and understanding from local team members and the business environment.

- Internal programs focusing on the celebration of various cultures represented by team members. A day-long immersion occurs at HQ with training, social activities (including local cuisine), external speakers as well as team members who represent these cultures.
- International competence training involving week long off-site training activities with the global project leader and team responsible for the global launch project.
- International meetings in various subsidiary locations and key markets for a day of local and regional product training, discussions, and team-building with a fun, interactive agenda.

Bibliography

Company overview, BMW Group: https://www.bmwgroup.com/en/company.html. Accessed on December 10, 2016.

Company profile, Mazda: http://www.mazda.com/en/about/profile/library/. Accessed on December 10, 2016.

Embracing Innovation: Case Study, "Inside high: Mazda's Seita Kanai on innovation and branding strategies", PwC: http://www.pwc.com/gx/en/services/advisory/consulting/revitalizing-corporate-japan/embracing-innovation-case-study-inside-high.html. Accessed on December 10, 2016.

Jensen, Karina R. Accelerating Global Product Innovation through Cross-cultural Collaboration Study, Report 2011.

Jensen, Karina R. Global Innovation and Collaboration Study, 2014.

Jensen, Karina R. Local Innovation and Collaboration Study, Asia Region, 2015.

Steep, Mike. "How to Create Innovation Cultures That Keep Working", Forbes Leadership Forum, Forbes, September 3, 2014: http://www.forbes.com/sites/forbesleadershipforum/2014/09/03/how-to-create-innovation-cultures-that-keep-working/#4c49381397ca.

Part III

Space: Creating an Environment for Inclusive Innovation

Vision
Leadership & Strategy

Space
Organizational Culture & Climate

Dialogue
Knowledge-sharing & Learning

6

Developing a Global Innovation Culture

Very pleased and enthusiastic about recent global launch results, a senior manager responsible for international products had enjoyed a successful career at a leading US multinational firm in the internet information services sector. Since he joined the company, it had enjoyed rapid international expansion and recruitment of a culturally diverse workforce. Describing the organization in a few words, he summarized the company as one with "worldwide reach, talent everywhere, an engineering focus and open offices in a majority of countries". In sustaining its start-up culture with a talented and diverse workforce, the company had succeeded in developing an environment that acknowledged contributions and local knowledge by employees from different regions. "There's a sense of empowerment at our company," the international product manager noted while adding that "A majority of people have worked on different projects in different countries and regions. You can also reach out from silos and contribute to other product groups. This empowerment allows for an innovation culture."

The senior manager explained that processes are both global and local in application where information is shared easily across teams and geographies. This approach makes it easier to drive projects across teams and locations. With full project transparency and local responsibility, collaboration takes place in live and virtual settings with geographically distributed teams. Optimizing a shared technical infrastructure and platform as well as informal team interactions, the manager explained "This process fosters entrepreneurial thinking and development with easy sharing of 'crazy' ideas."

© The Author(s) 2017
K.R. Jensen, *Leading Global Innovation*,
DOI 10.1007/978-3-319-53505-0_6

Yet there is more work ahead as the senior manager noted they have not yet found a fair and balanced communication process from concept to market. The future focus will be placed on more inclusion of teams working in remote locations around the world. Technology will serve a greater role for team communication, creativity, and knowledge-sharing. While structure is needed, he emphasized that "chaos needs to be part of the plan, otherwise there is no margin for creativity."

Nurturing an Inclusive Environment for Multicultural Organizations

Organizational culture serves an important role in fostering multicultural collaboration and innovation. A global innovation culture is necessary for creating a shared space with common values. It can also be viewed as a valuable strategic resource as it can provide a supportive structure and strategic motivation for global product innovation and launch, demanding further research and understanding (Calantone and Griffith 2007). Moreover, organizational culture is an important determinant for the climate of innovation as measured by the adequacy of resources, the encouragement and support of change and creativity and its impact upon strong and visionary leadership (Sarros et al. 2008). In order to meet the demands of a changing global marketplace, companies need to consider an organizational culture and climate that nurture and sustain innovation for culturally diverse teams and customers.

Organizations should also consider the global and local contexts that influence the development of common values and beliefs. Diversity and sharing, not conformity and protection, are acknowledged for creativity, innovation and cross-border collaboration for organizational cultures (Davenport et al. 2006). There is the need for balance in ensuring a unifying organizational culture that can also promote and optimize the knowledge diversity of local cultures and markets.

In developing global innovation capabilities, organizations can leverage cross-cultural collaboration and meet business objectives through increased market success worldwide. This means that international firms will need to optimize cultural diversity rather than manage cultural differences. Denial of cultural diversity has been shown to have a negative effect on innovation performance and project performance (Bouncken et al. 2008). Therefore, it is important to consider the role of organizational culture and its impact on multicultural collaboration and global team performance in developing new products and services.

In considering the critical elements for developing a global innovation culture within the organization, there are specific capabilities that include knowledge-sharing, worldwide team contribution, local responsiveness, formal/informal communication, and local resource commitment (Jensen 2011). The importance of worldwide team contribution is often mentioned, as noted by a global product director: "Our company is culturally diverse and we value differences where culture is an advantage." In addition, collaboration is often emphasized as a means to innovate as stated by a senior manager: "We emphasize collaboration everywhere with visibility for all regions." Many leaders spoke about the need to innovate by utilizing global and local knowledge as explained by one team leader: "It is important for our teams to work in both headquarters and local sites to develop and market global products."

Then what are the values that reflect a global innovation culture? Nurturing an environment for innovation in global organizations requires cultivation of three values: **cultural empathy**, **creativity**, and **collaboration** (Jensen 2011). *Cultural empathy* involves global teamwork and cultural diversity. Organizations need to ensure recruitment of a global talent pool in developing culturally diverse teams. This requires an awareness and openness to other cultures through international education, projects, exchanges, and collaboration worldwide.

Creativity demands idea generation and innovative thinking. The continuous development of new ideas along with free-form thinking is essential to an innovation focus. This requires both a creative and entrepreneurial spirit where adaptability and agility are important to initiating and realizing new ideas. An openness to risk-taking and potential failure is important for multicultural teams when experimenting with new solutions that involve ideation, rapid prototyping, and piloting.

Collaboration requires transparency and knowledge-sharing in order to build an inclusive culture. This involves collaboration with cross-functional and cross-cultural teams based in headquarters and subsidiaries. The use of frequent communication practices and tools with easy access to information and resources are critical factors to obtaining local market information. Organizations with strong innovation cultures optimize local market knowledge of team members through a focus on cultural empathy, creativity, and collaboration.

The Drivers of a Global Innovation Culture

The global innovation culture values represent an integration of organizational strengths from leading multinational firms. All of these organizations are active in recruiting global talent and promoting culturally diverse teams.

However, practices for developing cultural understanding and awareness vary from dedicated programs to limited resources. The creativity and innovation practices in organizations ranged from a company-wide framework supported by collaborative technologies to focused innovation teams with specific business objectives. The emphasis on collaboration and knowledge-sharing is highly dependent on the transparency and openness between team members located in HQ and subsidiaries. Some organizations have succeeded in developing global knowledge-sharing practices with continuous interactions between global HQ and local subsidiaries, while others face barriers to facilitating communication for geographically distributed teams. In order to provide more insights on developing a global innovation culture, let's take a closer look at values and practices for cultural empathy, collaboration and creativity.

When evaluating values that are most important to developing a global innovation culture, there are two priorities for organizations: 1) knowledge-sharing and exchange and 2) initiative and risk-taking. Additional values that received high priority included facilitating formal/informal communication, respecting and understanding cultural differences, and ensuring global and local responsiveness (Jensen 2012). There is an interesting relationship between priorities for a global innovation culture and actual organizational strengths, where execution and systematic project processes were considered the greatest strength by a majority of organizations (63%) followed by collaboration and visibility across cultures (40%). On the other hand, entrepreneurial initiative and risk-taking; idea generation and creativity as well as knowledge-sharing and networking received the lowest scores (Jensen 2012). Given the importance placed on knowledge-sharing and risk-taking, the values of cultural empathy, creativity, and collaboration have added significance in creating a global innovation culture (see Fig. 6.1).

Optimizing Diversity Through Cultural Empathy

An appreciation for cultural diversity and the value of a global talent pool is essential for orchestrating innovation around the world. Leaders and managers that can easily facilitate collaboration with culturally diverse teams have an advantage in ensuring international project success. Facilitating integration and collaboration while recognizing local talent contributes to greater inclusion and recognition across cultures and regions. As emphasized by a director of global product marketing: "We value cultural differences and we are a culturally diverse company that we view as an advantage".

Optimizing diversity requires increased cultural awareness and understanding through international training, rotational assignments, mobility, team leadership and cross-cultural learning. As explained by a senior manager responsible for global products: "We ensure diversity in our teams through a cross-cultural investment. This includes cultural awareness, education, and openness to other cultures through cultural theme events and team-building activities." In order to leverage local talent and knowledge, innovation is initiated locally and globally through shared practices. As noted by a senior manager responsible for international programs: "Our company reflects a global innovation culture through several processes designed to enable local input into global product planning and development."

The ability of cross-cultural teams to create trust, communicate, and develop strong relations can help facilitate cultural knowledge that drives the creation and implementation of new ideas. Cross-cultural interactions take place among social systems of different cultures and thereby constitute a common cross-cultural space. Cultural dimensions, cultural standards or

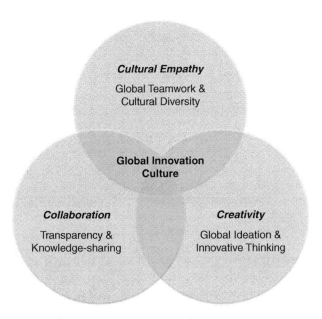

Fig. 6.1 Key values of a global innovation culture

interactions, and personality traits are related and determine behavior in cross-cultural interactions (Fink et al. 2006). In order to facilitate multicultural collaboration, an appreciation of cultural diversity and a capability for cultural empathy is necessary.

Innovative Thinking Through Creativity

In nurturing ideation and innovative thinking, creativity serves an important role in empowering teams to share ideas and to take new initiatives. In addition to fostering an entrepreneurial spirit, there needs to be an openness to innovation in order to develop creative thinking. There is also the pressure to become more agile, adaptable, and responsive to market needs. Many study participants noted that innovation was embedded in their company mission and values. An international management director emphasized that "we are actively rewarding and promoting innovative thinking – this is more often than not a global engagement and collaborative effort since so much is both product/process/business model related."

There is also the focus on developing behaviors for embracing risk and creativity. Organizations need to be more agile and responsive as expressed by a director of international product planning "Everyone is encouraged to think differently, to find competitive advantages across practices and regions." An openness to ideas and taking initiative creates the ability to connect local ideas with the global network for exchange and diffusion. This helps strengthen the connection to local customers and markets as emphasized by a director of global strategy: "We recognize local needs and consider every customer's norms and habits."

The development and sustainability of idea generation and co-creation requires an open environment for team creativity within the organization. Leaders set the tone for creativity and consequently develop an environment that determines to which degree organizational members generate creative work (Puccio et al. 2011). Creativity and innovation enable change within the organization through leadership that empowers team members to share and co-create knowledge.

> **Creating Collaborative Spaces Through Visual Communication**
>
> Visual communication in the form of images and illustrations can effectively create a common language for enhancing multicultural team collaboration. Visuals enable the expression of new ideas and concepts during ideation

sessions, in addition to creating new visions and strategies for planning sessions. The opportunity to use art and design as a process for creating global understanding and support across cultures is based upon the ability to encourage individual reflection and collaborative exchange. Leaders and managers can use art to facilitate knowledge-sharing and cross-cultural learning. There is an opportunity to evaluate and understand the interdependencies within global project collaboration. Some of the options to consider for your team include:

- Graphic facilitation
- Visualization
- Post-it idea generation
- Mind-mapping
- Prototyping
- Arts-based methods

The process of working with visual elements encourages dialogue and co-creation of the purpose and outcome needed to achieve successful results. Art and design are especially useful in bridging cultural gaps to facilitate sense-making and connection within a multicultural team. Visualization techniques such as graphic facilitation can be used in the development of a shared vision and strategy, helping global team leaders facilitate cross-cultural and cross-functional team interactions. There are several methods available from design thinking to arts-based to visual facilitation. Visual communication can bring valuable insights to the team when developing a collective vision and a creative space.

Transparency and Sharing Through Collaboration

Transparency and knowledge-sharing help nurture an environment for collaboration amongst geographically distributed teams. Navigating and adapting to changing project needs requires global and local collaboration between teams in headquarters and subsidiaries. This requires transparency and an open environment that facilitate access to information and resources globally. It requires a collaborative team focus and interactions across regions. There is also the need for easy adaptation and integration in order to share best practices and ensure learning between various cultures and regions. The vice president of an information services firm stated that "our internal structure leverages talent around the globe to create efficient teams that deliver on specific targets and initiatives…we frequently move people across geographies to create understanding, appreciation, and awareness of the different cultures."

An open and flexible environment increases communication and knowledge-sharing with teams around the world. The structure needs to support a network-centric organization with reduced layers and decision-making mechanisms. The international marketing director of a multinational firm in the personal care industry explained that "we promote an informal and flexible culture that is open to adaptation, styles, motivation, and experiences." As demonstrated in the Lenovo case on optimizing cultural diversity, organizations that enable communication transparency can facilitate input by everyone in the organization, including cross-functional and cross-cultural teams at global and local management levels. This facilitates networking capabilities for team interactions during onsite meetings as well as online collaboration across geographies.

> **Performance Excellence Through Cultural Diversity at Lenovo**
>
> Lenovo is a global personal technology company with a worldwide presence that includes 55,000 employees and customers in 160+ countries. The company started in Hong Kong in 1988, growing into the largest PC vendor in China and the world with the acquisition of IBM's PC Division. Today it offers a global product portfolio that includes tablets, smartphones and smart TVs. The company has developed an organizational culture that is focused on break-through innovations and performance excellence through the Lenovo Way values and message: We do what we say and we own what we do.
>
> The company has successfully integrated an Eastern culture with Western practices to become a multicultural organization. There is less hierarchy and more focus on the organizational network with regional headquarters and global responsibilities. Lenovo values cultural diversity and an entrepreneurial spirit where there's a sense of inclusion and respect. Cross-cultural and cross-functional teams enjoy systematic project collaboration with a strong execution focus and high performance. International mobility is emphasized throughout the organization with overseas assignments, travel, exchanges and global team collaboration. Cross-cultural training courses are also offered in order to effectively manage across cultures. In developing talent worldwide, culturally diverse management teams are found across international locations which further strengthens Lenovo's global innovation environment.
>
> Employees are encouraged to contribute ideas and knowledge to accelerate innovation throughout the company. In addition to a dedicated project portal, there is a knowledge platform with social media tools and blogs where team members can share their perspectives. Regular meetings are focused on knowledge-sharing, success stories, and lessons learned. In addition to an internal Wikipedia, cross-functional teams can use queries to search global and local knowledge in order to exchange and communicate the latest information across geographies. Creating a global innovation culture through cultural

empathy and collaboration, Lenovo is able to strengthen its customer-centric focus in diverse international markets.

Sources: About Lenovo page, Lenovo: http://www.lenovo.com/lenovo/us/en/index.shtml?footer-id=our_company. Accessed on November 2, 2016.

Jensen, Karina R. Global Innovation and Collaboration Study, 2014.

Weber, Lauren. "Changing Corporate Culture is Hard: Here's How Lenovo Did It", The Wallstreet Journal, August 25, 2014.

Optimizing a Global Innovation Culture Through Organizational Routines

The organizational culture serves an important role in shaping an environment that is conducive to multicultural innovation and collaboration. The study findings presented in Chapter 4 showed that national culture affects knowledge-sharing behaviors within the organization. However, the minority (12%) that found national culture did not affect knowledge-sharing indicated that organizational culture had helped integrate cultural differences by creating common values for teams worldwide. If team members support cultural empathy, creativity, and collaboration, there is the opportunity to embrace organizational values that support and integrate different cultural views and practices.

In exploring values for developing a global innovation culture, there are three distinct themes that have been presented: cultural empathy through an emphasis on global teamwork and appreciation for cultural diversity; creativity through ideation and innovative thinking; and collaboration through communication transparency and knowledge-sharing. Values are important, however, the mindset and routines are necessary to create a

Table 6.1 Organizational routines for a global innovation culture

Organizational Routines
• Cross-cultural team interactions
• Shared work space
• Effective project process
• Collaboration and communication tools
• Inclusive team leadership

dynamic organizational culture. What are organizations doing to make this happen? As shown in Table 6.1, they are using specific organizational routines for shaping a global innovation culture that include opportunities for **cross-cultural team interactions, a shared work space, an effective project process, collaboration and communication tools, and inclusive leadership** for all team members (Jensen 2014). Organizations are finding it necessary to create the time, space and resources dedicated to knowledge-sharing and cross-cultural learning for sustaining innovation around the world (Table 6.1).

Cross-cultural Team Interactions

Although the world is becoming more digitally connected, there is still a greater preference for face-to-face communication in managing effective project collaboration. Executives and senior managers leading global innovation projects find there is too much focus on web communication tools and technologies and not enough emphasis on international mobility (Jensen 2014). An overreliance on geographically distributed virtual teams and use of communication technologies have resulted in increased collaboration issues and cultural tensions for many multinational firms.

In order to build trust and cultural understanding, regular international travel and meetings between teams based in headquarters and subsidiaries are important for team collaboration. However, international mobility also depends upon leadership support and travel budget allocation. International forums and planning meetings organized for teams worldwide offer an opportunity to learn about local market practices and to collaborate on new innovation projects. Most of all, increased time spent in the field visiting key regions and countries allow improved understanding of local customer preferences and market practices.

Shared Work Space

Since there is a great need to develop more opportunities for face-to-face interactions, a common space for the global project leader and cross-cultural team members can facilitate dialogue and exchange for global innovation, especially in the front end. Idea-pooling workshops or dedicated meeting spaces for ideation and planning help accelerate project collaboration. It's an effective way to optimize multicultural innovation through on site sharing and validation of new ideas from diverse markets. On site work spaces combined with a technology platform for knowledge-sharing initiate and sustain communication for critical project phases. This creates more time, space, and freedom for global team members to create new ideas and concepts that respond to international market opportunities.

Collaboration and Communication Tools

In addition to an emphasis on live collaboration and interaction, study participants emphasized the need for a dedicated technology platform for ideation and knowledge-sharing. Social media and web communication tools that facilitate team collaboration and spontaneous knowledge-sharing are also popular. However, the most in demand tools are those that simulate human interaction through visual connections, especially telepresence and video conference. The technology tools and team communication should be matched to the project collaboration process. In general, the global project team requires an internal social network or knowledge platform to make connections and share ideas, in addition to a project portal with flexible communication tools for moving from idea generation to implementation phases.

Effective Project Process

A dedicated and integrated project system is essential to connecting team members throughout global innovation cycle. This places greater emphasis on project leadership that can co-create a clear strategy and plan as well as facilitate the project process. Organizational communication and learning resources that can strengthen project collaboration with cross-cultural teams involve global team communication, knowledge-sharing structure, online communication technologies and tools, and cross-cultural team learning. There's also the need to consider incentives, time, structure, and processes available for ensuring sufficient knowledge-sharing.

Inclusive Team Leadership

An environment that is conducive to global innovation is shaped by the inclusive mindset of its leadership and team members. This behavior is developed through everyday practices that support open and receptive communication within teams. Relationship-building develops increased cultural understanding and the capacity to learn through knowledge-sharing. It requires a leadership team that is willing and able to enable teams through trust, commitment, and support. The organizational environment and leadership are essential components for sustained global innovation and performance in cross-cultural contexts.

Research has shown that a global innovation culture places particular emphasis on values for cultural empathy, creativity, and collaboration (Jensen 2014). As shown in the Siemens Convergence Creators case, these values are shaped by global teamwork and cultural diversity, ideation and innovative thinking, transparency and knowledge-sharing. Moreover, they are enabled through routines that include an inclusive mindset, cross-cultural team interactions, a shared work space, an effective project process, collaboration and communication tools. As discussed in this chapter, the foundation for global innovation and collaboration involves the development of an inclusive organizational culture.

> **Siemens Convergence Creators: Global Knowledge Exchange and Collaboration as Key to Business Transformation**
>
> Siemens Convergence Creators delivers highly innovative telecommunication, media technology and satellite products and solutions to customers in more than 70 countries. The company operates sites across multiple geographies in Asia, Europe, the Middle East, and North America. The regional offices are located in Austria, China, Croatia, the Czech Republic, Germany, India, Romania, Slovakia, Saudi Arabia, the United Arab Emirates, and the US.
>
> Starting in 1961, the predecessor organizations delivered outsourced R&D services for five prosperous decades. Facing fierce global competition, it became clear that offering commoditized services would not secure sustainable business success. Accordingly, the organization's strategy was significantly re-defined in 2010, focusing on own products and solutions providing specific value and thus leading to a distinctive market positioning. Continuous innovation laid the foundations for this substantial transformation: Five years later around 80% of sales are related to new portfolio elements.
>
> Although most of Siemens Convergence Creators' portfolio elements are managed at the Vienna headquarters, smooth and close international co-operation is crucial for the company's success. In general, most projects and product development are performed by geographically distributed teams. Especially for innovation, leveraging the full creative potential of all employees, no matter where they are located, is of utmost importance.
>
> Hence, the key question is: What can be done to create an ecosystem in which global collaboration takes place as if no distance would separate teams and employees?

Let us have a closer look at some hurdles that may stand in the way. Geographical distance affects cooperation in many ways: These include psychological aspects such as team and trust building as well as asynchronous working hours in different time zones. The lack of spontaneous face-to-face communication options, language as well as cultural issues restrain extensive, unhindered interaction.

Acting from a mainly local perspective may also lead to an attitude of internal competition, which is a mixed blessing: On the one hand it may stimulate a local high-performance culture, on the other hand it may also obstruct teaming up on a global level to achieve the company's goals.

The collaboration and innovation ecosystem at Siemens Convergence Creators addresses these issues from a holistic organizational perspective, including cultural topics as well as IT tools, knowledge sharing and processes. Some examples:

- To facilitate global communication on an equal basis, English is used as the corporate language, although the language of the largest employee group and at the Vienna headquarters is German. Accordingly, all processes and regulations are written in English and laid down in one globally valid web-based process handbook, keeping local amendments to a minimum. This ensures the globally same understanding of how things shall be done, and thus smooth interworking and consistent quality of work across all borders and boundaries.
- The same philosophy is reflected on the intranet employee portal, offering news and access to information resources to all employees. Again, this portal is globally available as a single point of entry. While most content is provided by the headquarters, information in the local language is amended in each country and displayed along with the global information.
- Innovation topics play an important role in global communication; a specific sub-site at the employee portal offers information on the company's innovation strategy as well as possibilities to submit innovation and improvement ideas or to file a patent. All submissions, no matter where they originate, are received and processed by the headquarters' innovation management team. This direct access ensures that all submissions are treated in an unbiased way, also bypassing middle managers who may not actively support the employee's ideas.
- The headquarters' innovation management team also ensures that promising ideas are adopted by business units – if this is not possible, decisions are made directly on the top management level. This may result in setting up an own project team mandated to further elaborate the idea.

A key prerequisite for fruitful idea generation is knowledge about technological and market trends, customers' needs and competitors' offerings. Of course knowledge sharing is most effectively done personally. Hence, training on innovation theory and methodologies as well as idea creation workshops are offered in most countries. Web-based formats complement personal attendance events, supporting knowledge sharing at little time and cost expenditures. The "Innovation Lounge" is a mixed format, in which invited internal and external speakers share information on specific topics. These events are performed at headquarters as well as broadcasted via web to all locations. A recently introduced format combines focus and modularity: The "Short Cuts" are 20-minute live webinars on single topics such as "Patent research" or "Blue Ocean Strategy", "License models" or "Disruptive innovations", offering additional 10 minutes for Q&A. An innovation in itself, this again globally offered format has met with great enthusiasm.

The well-functioning international collaboration of distributed teams at Siemens Convergence Creators is based on a long tradition of cooperation. This culture has been strengthened in the past years by introducing a globally valid and easily accessible process handbook and by offering information via a common intranet platform. Yet, the availability and use of teleconferences may lead to the false conclusion that personal presence is no longer needed. Web-based tools and formats support the exchange of information and ideas, but experience gained at Siemens Convergence Creators clearly shows that virtual meetings cannot fully substitute personal face-to-face communication. The major cultural challenge remains to ensure bi-directional personal communication and mutual appreciation – the basis of cross-cultural collaboration leading to competitive advantages. Thus finding the right balance between personal and web-based meetings is a continuous task.

Author: Hans-Jürgen August, Vice President Innovation and Quality Management, Siemens Convergence Creators, Autokaderstr. 29, 1210 Vienna, Austria.

Practices that Nurture and Sustain a Global Innovation Culture

- Brief internal meetings where members from various countries and functions are invited to exchange practices via video or teleconference with various regions during the same time period.
- Idea pooling workshops with team members from HQ and local subsidiaries for project collaboration at the front end phase.
- Networking events such as cross-team meetings where two members from a local subsidiary run a workshop for team members based in HQ concerning a topic of particular interest, followed by a social activity.
- Holding global and local customer forums and brainstorming sessions with cross-cultural team members in facilitating ideation and validation of new concepts.
- Annual company conferences dedicated to global innovation for employees and team members worldwide where everyone submits ideas.
- Shared reward systems where teams are directly compensated for project results.

Bibliography

Bouncken, Ricarda B., Ratzman, Martin and Winkler, Vivian A. "Cross-Cultural Innovation Teams: Effects of Four Types of Attitudes Towardds Diversity. *Journal of International Business Strategy*, 8–2 (2008): 26–36.

Calantone, Roger J. and Griffith, David A. From the Special Issue Editors: Challenges and Opportunities in the Field of Global Product Launch. *The Journal of Product Innovation Management*, 24 (2007): 414–418.

Davenport, Thomas H., Leibold, Marius and Voelpel, Sven. *Strategic Management in the Innovation Economy: Strategy Approaches and Tools for Dynamic Innovation Capabilities.* Erlangen, Germany: Publicis and Wiley, 2006.

Fink et al. "Understanding Cross-Cultural Management Interactions". *International Studies of Management and Organization*, 7(2006): 38–60.

Jensen, Karina R. Accelerating Global Product Innovation through Cross-cultural Collaboration Study, Report 2011.

Jensen, Karina R. Global Innovation and Collaboration Study, 2014.

Jensen, Karina R. Global Product Innovation and Cross-cultural Collaboration Online Survey, 2012.

Puccio, Gerard J. et al. *Creative Leadership: Skills that Drive Change*, Thousand Oaks, CA: Sage Publications, 2011.

Sarros, James C., Cooper, Brian K. and Santora, Joseph C. "Building a Climate for Innovation through Transformational Leadership and Organizational Culture", *Journal of Leadership and Organizational Studies*, Nov 15–2 (2008): 145–158.

7

Creating an Inclusive Innovation Climate

Scrambling to keep up with competition as the company had lost market share to key competitors in both mature and emerging markets, a senior director for an international telecommunications firm based in Europe found herself in a position of trying to determine markets where the company could make the largest impact. There was a need to align product creation and manufacturing capabilities with marketing in key geographic locations. In addition, she needed close collaboration with local teams to capture up to date information on customers, points of sales, and real time understanding of market activities in order to link planning to execution. "There's always a lack of time, we need more focus and time to gather information and knowledge for each phase," she noted with a frustrated tone.

There was information overload within the organization where some regions were screaming louder and making demands to achieve greater influence in planning and resource allocation. This made it challenging to prioritize markets and determine needs for local market success. It was difficult to access local knowledge and move information through the appropriate communication channels in the organization. Local markets were feeling the pain in terms of competitive threats, a shortage of supply, and regional market dynamics. It placed teams in a short-term focus of market exploitation and did not allow them to innovate through market exploration in order to create new opportunities.

In order to improve the team innovation climate, the global planning director decided to develop a more collaborative approach to strategic planning and execution for a new product line. "It's important for the team to understand when and how decisions are made…you should be open to this process," she emphasized. She created a process where everyone's feedback could be captured through frequent face-to-face meetings as well as use of the internal communication platform and online tools. "What is rich in cross-cultural team communication is that you look at the global perspective and get views from different cultures which provides interesting perspectives," she paused, then added "the key is contextual sensitivity and managing teams through communication."

The Global Innovation Cycle and Project Process

The global innovation cycle demands a multi-faceted, multi-functional, and multi-cultural approach to planning and execution. Organizations are increasingly focused on accelerating time to market through reduced product development cycles and increased process efficiency. Yet time is needed for customer insights, market exploration, cultural understanding, and creative thinking. There has also been great emphasis on particular stages such as Ideation, R&D, or New Product Development, rather than attention to the full innovation cycle, from concept to market. Although the launch of new products remains a core business priority for international organizations, there are few project and team process models to guide leaders and managers working in international and multicultural environments.

While standardized models are available for product management and project management, there are few that address the operational and internationalization needs for the conception and introduction of new products to international markets. The global innovation project requires special considerations for the three key stages of Front End Innovation (FEI), New Product Development (NPD), and Go-to-Market/commercialization (GTM). There are multicultural collaboration needs during the global innovation cycle that involve local knowledge-sharing and communication. Planning and execution have to address local customer and market preferences, including features, languages, images, and messages. These are critical stages in preparing new concepts and content for key markets worldwide.

Multicultural Project Collaboration, from Planning to Execution

Managing across functions and cultures while leading an innovation project requires a team climate conducive to collaboration. A solid project methodology and communication process serve as the framework. The few global innovation management models that exist are often adapted from new product development or project management disciplines. While universal and standardized models have been introduced for product management and project management, from stage gate to agile, they rarely address the special cognitive needs of the global innovation cycle such as social networking, knowledge-sharing, and learning within culturally diverse and digitally connected business environments.

As shown in Fig. 7.1, the global innovation cycle is a comprehensive process that moves from creation to strategic planning to global launch execution. When creating and introducing new concepts to international markets, there are five key phases that are followed by the global project team which include (1) Creation for discovery and ideation, (2) Strategic Planning for market analysis, business case development and a launch plan, (3) Validation of the concept with local teams and customers, (4) Execution

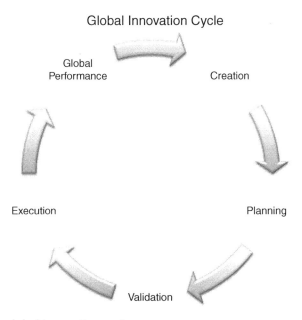

Fig. 7.1 The global innovation cycle

through new product development and go-to-market activities leading up to (5) Global Launch and delivery of the final concept. While the innovation cycle includes these key phases, it is not necessarily circular or linear (as shown in Figs. 7.1 and 7.2) where some phases may require earlier and parallel timelines as well as an interative process depending on concept creation and go-to-market needs.

Close collaboration and communication are required from planning to execution to ensure concepts and programs are suitable for international markets. The international project lead and functional teams at headquarters work closely with local product, marketing and sales teams at the subsidiary level. Product, marketing, and sales readiness should meet pre-defined criteria during launch preparation in order to be set for the worldwide announcement.

The creation, planning, and delivery of new concepts demand the involvement of cross-functional, cross-cultural, and executive teams. These stakeholders serve an integral role in project success. The leader and core team drive the global innovation project. This requires an international framework and scope, project process, communication and collaboration tools. The global project also includes cross-functional teams responsible for product design, product development, operations, marketing, and sales. In addition, regional or local marketing and sales teams are actively involved and responsible for planning and implementation. As always, there is the need for sponsorship by the executive team who provides the authority and influence for decision-making and approval of key planning and execution phases for the global project process.

As shown in Fig. 7.2, the main stages that drive the global innovation project process include Front End Innovation (FEI), New Product Development (NPD) and Go-to-Market (GTM). Supporting the global innovation cycle, the five project phases that link the three innovation stages include **Creation, Planning, Validation, Execution, and Global Launch**. *Creation* incorporates the preliminary phase of local market understanding and discovery which involves market insights, need identification, and articulation. The *Strategic Planning* phase moves from local to global learning and integration in sharing market intelligence and research for evaluating and developing the business case. The *Validation* phase is critical to identifying and designing internationalization needs through testing and feedback. And the *Execution* phase determines the extent of localization for features through concept development and testing while marketing requires attention to positioning, messaging, pricing, content and media. This leads to international marketing and sales readiness for global to local launch results.

An organization's particular focus on radical or incremental innovation may involve an additional step or two depending on the need for more

Fig. 7.2 The global innovation project process

exploration and discovery as well as testing and validation with local teams and customers. The phases that are most critical for successful concept development and market introduction involve local market research and analysis, concept creation, effective planning, and execution for go-to-market needs.

A roadmap should always guide your team throughout key stages and phases of the global innovation cycle. As shown in Fig. 7.3, the execution phase involving global launches requires special attention to strategic planning, internationalization and localization of the product/service concept as well as messaging and marketing content, and international sales readiness through local training and support. Internationalization or "i18n" is an essential process in the design and development of the new concept as well as marketing content in order to facilitate localization in various cultures, regions, and languages. Localization or "l10n" is more focused on the adaptation of the product, application, or content to meet language, cultural, and local market requirements (W3C 2016).

The project management process for global product innovation requires special consideration for idea generation, conversion, and diffusion of new concepts. Adaptation and customization to local customer preferences is required since the global product launch involves consideration of diverse markets, cultures, and languages. There will be the demands of prioritization

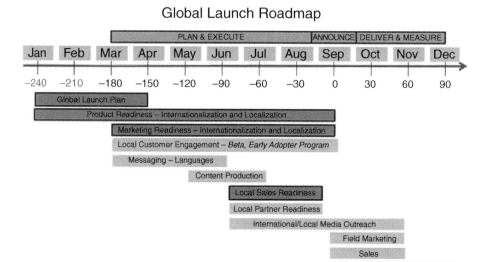

Fig. 7.3 Global launch road map and key milestones

for markets and resource allocation, sufficient training and support. Planning and execution activities need to consider local market adaptation needs, including features, languages, images, and messages. Internationalization in concept design and development enables localization and adaptation to language and cultural needs. These are critical steps in preparing new concepts and content for key markets worldwide. In order to ensure cross-functional and cross-cultural teams stay on track, make sure that everyone shares the same roadmap (see Fig. 7.3).

In comparing the planning and execution phases, there are distinct differences in the type of knowledge sought and shared by the project leader and the geographically distributed team. The planning phase requires local market, customer, and product knowledge directly related to the creation and validation of the product concept. Critical internationalization and localization requirements need to be determined during this phase such as customer preferences, product features, pricing, marketing and sales resources.

On the other hand, the execution phase requires more specific knowledge concerning local marketing and sales content and support. This phase also involves final localization requirements and customer engagement for concept validation. The planning phase determines the key criteria for the success of the global product launch project while the execution phase ensures the development and delivery of the product, marketing, and sales content needed for the worldwide introduction.

Communication in Complex, Multicultural Team Contexts

The dynamic and complex process of leading global innovation projects with cross-cultural and cross-functional teams in multiple locations is not an easy role. Yet, it is an exciting, constantly changing, and rewarding learning adventure that brings rich cultural insights and international market opportunities. Team leadership and communication are considered the top factors in facilitating multicultural collaboration during the global innovation project (Jensen 2014). Leaders require effective communication skills in order to meet the challenges of uniting geographically distributed teams, facilitating knowledge-sharing and collaboration as well as orchestrating the project process worldwide.

The project collaboration process is central to team engagement since individual contribution is often linked to motivation and trust. There is a need for effective communication, knowledge-sharing and feedback processes throughout every phase of the project process – Creation, Planning, Validation, and Execution. Trust-building, knowledge-sharing, and team engagement are critical in the early phases of Front End Innovation, from creation to planning and validation. Leaders and managers need to facilitate and integrate communication and knowledge-sharing throughout the project collaboration process. As shown in Fig. 7.4, each phase brings a particular need for sharing, learning, and integration.

The *Creation* phase provides the opportunity to learn and discover through cultural immersions and travel between geographies where teams can provide their insights and create concepts. There are both global (HQ management teams) and local (all subsidiaries) or glocal knowledge-sharing and ideation.

The *Planning* phase allows for strategic co-creation where regional and international forums are organized in the various geographies to share and integrate knowledge that will shape the global strategy.

Moving into the *Validation* phase, there is the need to test the proposed concept with local teams and customers in order to ensure internationalization

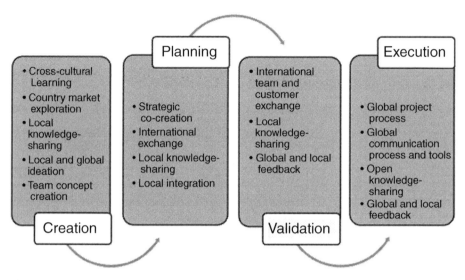

Fig. 7.4 Multicultural collaboration project process

and localization. This requires knowledge-sharing and feedback from key markets and regions. Depending on whether the innovation is incremental or radical, it may require validation prior to and after strategic planning. Finally, the team is ready for the *Execution* phase where it's essential to design an integrated project and communication process for knowledge-sharing and communication built into milestones for international concept, market, and sales readiness.

Development of an Inclusive Innovation Climate

A global innovation culture provides the framework while the innovation climate enables the network and connection to international customers and markets. Climate can be viewed as recurring patterns of behavior, attitudes, and feelings that characterize work life for teams; facilitation and leadership establishes, nourishes, and maintains a climate appropriate for the team to succeed (Isaksen and Lauer 2002, Isaksen and Akkermans 2011). Organizational climate is closely tied to the experience and interactions of individuals and teams within the organization. Motivation is often developed from social conditions and encouragement in the form of expertise and creativity which involves collaboration, diversity, support for creativity, and acceptance of failures (Amabile 1997, Amabile and

Khaire 2008). Organizational practices are increasingly focusing on optimizing cultural diversity through collaboration, creativity, risk-taking, and knowledge-sharing.

In order to create the optimal team environment for nurturing multicultural innovation and collaboration, four values should be considered for developing a global innovation climate: **market responsiveness, entrepreneurial spirit, global team transparency, and execution efficiency** (Jensen 2014). Fig. 7.5 shows how the team climate enhances a global innovation culture. *Market responsiveness* represents a strong focus on customer and market-orientation for leaders and their teams. *Entrepreneurial spirit* refers to the ability to take risks, to quickly embrace innovative ideas, and the willingness to invest in new initiatives. *Global team transparency* and *efficient execution* show a strong link to collaboration due to the need for knowledge-sharing using a structured and disciplined process. Team climate is an integral element of organizational culture when nurturing creativity and innovation within companies.

Market responsiveness is expressed in terms of a customer-driven focus with adaptability and agility for changing market needs. There is increased pressure to understand and listen to customer needs in regional and local markets around the world. A customer orientation demands more time and attention to cultivating relationships and building loyalty. Due to the fast-paced and

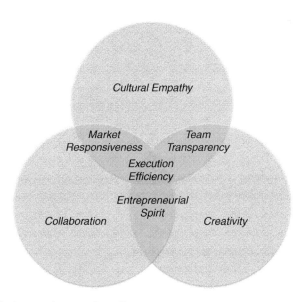

Fig. 7.5 Global team innovation climate

changing nature of international markets, there is an increased need for market responsiveness in being flexible, quick, and adaptable to challenges and solutions for international customers. However, organizations are also challenged by a lack of market responsiveness and customer focus. The inability to respond to customer and market demands can be hampered by the organization's administrative layers and slow decision-making processes. A lack of customer-orientation and attention to local market requirements can result in the inability of the organization to respond and provide the appropriate market solutions.

Global team transparency is closely linked to collaboration within a global innovation culture. There is an openness and ability to share and network across functions and cultures. This process involves participative decision-making and a consensus-driven communication style when facilitating multicultural collaboration for geographically distributed teams. As noted by a global program director: "We have very multicultural and geographically distributed teams. They are well recognized in the organization and deliver key innovations with the means and the resources." Teams require sufficient space, resources, and tools to optimize culturally diverse talent through collaboration and visibility.

On the other hand, there is also the argument there may be too much collaboration for certain project activities where group members take a long time to arrive at a decision. If collaboration is needed at every step of the launch process, it could slow the go-to-market activities. Yet, a lack of team engagement and visibility can decrease collaboration. Team leaders need to facilitate knowledge-sharing and communication and determine the level of engagement of cross-functional and cross-regional teams in the FEI, NPD, and GTM phases. Organizations are still struggling to achieve effective multicultural collaboration with leadership and team behaviors that support an innovation climate of transparency, responsiveness, initiative, and execution efficiency (Jensen 2014).

Entrepreneurial spirit is closely linked to the organizational culture value of creativity. It is about the ability to take risks and embrace new ideas, including the willingness to initiate and execute new solutions. It's also an openness to creativity and innovation in order to develop new opportunities within the organization. A new product management director explained that "we have lots of ideas and the benefit of an international market presence, we need to leverage international markets. For example, in Asia there is more openness to risk, more ideas, and more opportunities." The focus is increasingly placed on entrepreneurial or intrapreneurial initiatives that result in the generation of new ideas. As emphasized by a senior program manager: "As a culture and mindset, we focus on disruptive innovation and intrapreneurial skills."

On the other hand, entrepreneurial spirit can be a challenge in organizations where there is a lack of risk-taking due to a conservative organizational culture with administrative layers that can challenge innovation opportunities. As described by a senior global planning manager: "We take advantage of different initiatives in our organization but there's the complexity of layers. It takes longer to make decisions and we could be stopped by other issues in layers, so changes are needed." There is also the issue of placing too much emphasis on incremental improvements instead of placing the focus on disruptive innovation for new ideas. This requires leadership that can facilitate more idea generation and creativity in order to evolve the innovation process with teams throughout the global network. A global product manager complained that "we keep innovation in HQ for control of process. Since it doesn't happen at the subsidiary level, we miss out on creative ideas and opportunities to incubate."

Execution efficiency brings a structured and disciplined process that is beneficial for launching new concepts to international markets. A structured process orientation that ensures timely execution can be an advantage when responding to international market demands. As explained by a global product marketing director: "We need to be structured and disciplined due to the product design and manufacturing process. We have to stick with the timelines and we need to ensure that every year we have a new product or solution." Efficiency and prioritization of resources serve an important role for large scale projects such as new product introductions.

However, a solid framework is needed to move the product from concept to execution which may also require a flexible structure. If there is too much structure, the team can suffer the challenges of a rigid and complex process. Standardized and structured processes can slow NPD and GTM activities when selecting and executing on new ideas. As expressed by a Vice President of Product Marketing: "The structured, disciplined approach works well but it limits out of the box thinking." While there's the administrative burden of maintaining existing products, organizations need to keep their sights on new market opportunities and shifting consumer preferences in key countries and regions.

A global innovation climate helps develop a team environment conducive to multicultural collaboration. Market responsiveness demonstrates the need to be more customer and market-driven while demonstrating adaptability and agility to local market needs. A growing focus on entrepreneurial spirit is linked to creativity and the ability to generate ideas while taking risks and showing initiative. Global team transparency supports openness and knowledge-sharing in facilitating multicultural team collaboration. It's an appreciation for culturally diverse talent and knowledge where interaction and sharing is encouraged. However, collaboration also supports execution efficiency in order to ensure timely market response. This requires a structured and disciplined approach with an effective project process. Market responsiveness, global team transparency, entrepreneurial spirit, and execution efficiency determine the strength of an organization's team innovation climate.

Enabling Local Connections to Customers and Markets

The planning phase always requires local and regional access to market, customer, and product knowledge. There is a need to understand local market potential by examining trends, size, growth, and competition factors. There's also the necessity to understand the customer profile, preferences, needs, and expectations in developing a suitable product offering. In order to evaluate the launch plan, the global project leader also needs to determine product feature localization, pricing, and resource needs for marketing and sales activities. Finally, there is the need to assess financial resource allocation dependent upon budget needs in relation to forecasted revenue for the local market.

There is a constant alignment between the project leader in HQ and team members in local markets in order to ensure the product strategy meets local market expectations. As noted by a global product management director "We need to check if decisions made at HQ make sense with local market expectations. There could be business and technological constraints; we need to see how the new product will fit with business habits and its compatibility with consumer expectations and specificities." Specific and updated market information is critical to effective planning and execution. Knowledge about the customer and the market potential provides improved insights to strategic planning needs. A senior manager explained that "We need to have product development phase 0 ready and show this is a marketing opportunity and the market is profitable based upon these facts and assumptions."

There is the growing need to build a business case in order to demonstrate market potential and to obtain sufficient budget allocation. As noted by a senior manager, "We need to get the business case and target customers/users in challenging our own market analysis from HQ, it's bottom up vs. top down business planning." Thus, market insight combined with a sound business case can improve alignment between the project leader in HQ and the geographically distributed teams in local subsidiaries.

The planning phase also demands consideration of the content and the process of knowledge-sharing. In-depth investigation of customer requirements may be needed for product internationalization and localization. A director of product management noted that "very specific market needs are not a given such as features needed locally. For example markets that have specific needs due to climate or fuel criteria. Thus, you need to know the requirements from a technical view as well as local regulations due to test requirements." Time spent in local markets for discovery and analysis will uncover key criteria for ensuring effective project collaboration as well as a successful market introduction.

It's often necessary to understand customers' user preferences and their relationship to the products used. This demands access to markets and customers through local team members involved in the global launch project. A senior manager emphasized the importance of "access to the local market, know how customers use particular concepts, what are their wants, needs, and what they envision in terms of the kinds of options desirable. How can the product be used to improve life and productivity?" This kind of information can usually be obtained through the interactions between the global project leader and team members based in local markets. Such interactions demand frequent and consistent communication through knowledge-sharing. As explained by a product marketing director, "The most important information is resolved through active participation. We need more collaboratively structured planning, it's mostly top down now…most important, it needs to be an open, collaborative, and trusted process.

When seeking local market insights, marketing related information is sought for positioning, messaging, and localization of marketing content. There's also customer and sales information that is especially important for ensuring local sales support and training as well as partner support. Most important, there's customer knowledge and validation for gathering feedback, references, and ensuring support concerning the new product. There's also the need to check on final localization needs for products and marketing content should additional changes be required prior to the worldwide launch date.

Marketing and sales serve an active role during the execution phase where positioning and messaging are developed for international markets. A clear and unique value needs to be developed as a common theme for interpretation and translation to local markets. As stated by a senior manager, "the most critical information is about messaging. If you consider the value proposition, the value pillars, mix of messaging, media mix, and timing (the actual launch date)". Then there's the sales content and training that is needed to ensure the sales team has sufficient knowledge and readiness to sell into international markets. As noted by a senior manager, "There needs to be capability on execution, where you have resources lined up, trained, informed, and ready. A lot of education is needed in having marketing ready for international customer support and partner support." Customer intimacy places a focus on local market demands and the drivers that determine customer preferences. As emphasized by a director, "There's customer feedback for feature needs and requirements, as well as marketing material needed."

The go-to-market and execution phase ensures launch preparation from market awareness to sales readiness to customer engagement. There is the constant concern for sufficient communication and information. If a company succeeds in capturing positive, local customer references prior to the product launch, there is an opportunity to develop more persuasive marketing and sales tools. As emphasized by a vice president of product marketing: "We need to improve the international customer reference plan which shows adoption and falls between product development, marketing, and sales… we're good at recruitment, but need to foster a relationship that yields a visible reference." Attention should be placed on access to local customer references and information that can assist regional sales teams.

From Global to Local Solutions: Wipro's Collaboration Focus

Wipro Ltd is a global information technology, consulting and business process services company with 170,000+ workforce serving clients in 175+ cities across 6 continents. The company has an engineering DNA with deep technological expertise. Its 55+ dedicated emerging technologies "Centers of Excellence" enable it to harness the latest technology for delivering business capability to clients. Wipro works as a co-innovator to businesses across 22 industry verticals, identifying new growth opportunities and facilitating their foray into new sectors and markets. This enables Wipro to draw insights across industries, processes and technologies.

The company has been an early user of internal and external communication and collaboration platforms in order to nurture ground-up innovation. Wipro TopGear, its internal crowdsourcing platform provides virtual and physical environments for its 170,000+ global employees to discuss ideas, problems and create technological solutions. So far, TopGear has been instrumental in cross pollinating ideas that have been at the core of solving over a hundred problems. The theme based idea challenges on the Wipro TopGear platform underline Wipro's collaborative work culture.

Wipro's crowdsourcing vision took a quantum leap in 2016 with the acquisition of TopCoder, a 15-year-old innovative collaboration platform. Over the years, TopCoder has refined the art of solving problems, coding software, using ratings and attracting highly skilled members. The global TopCoder community has so far solved 41,000 challenges placed before it by over 500 customers. The platform is a testament to the power of multi-cultural collaboration that can successfully germinate and nurture solutions to complex problems.

More recently, Wipro has systematically launched global co-creation workspaces called "Digital Pods". These pods combine problem-solving skills from technology, development, engineering, strategy and design teams. Here, regular workshops are held in rapid-fire 24-hour sessions to develop new, disruptive digital solutions. The pods, distributed across the globe, bring together multidisciplinary teams from customers and providers, stressing a no-shore approach to synthesizing ideas and skills across domains, skills, cultures and geographies and to rapidly catalyze breakthrough solutions in insurance, banking, manufacturing and utilities.

Wipro's DNA of collaboration and its technological platforms allow it to identify and deliver innovative solutions that meet local customer and market needs, as demonstrated by the new concepts it has created in mobile health care. The company has developed two low-cost wearable devices at its India research and development facility for mobile health care. Both devices have been developed under the Wipro's Assure Health™ software platform. Assure Health™ combines mobile technology, advanced Body Area Networks (BAN), unobtrusive wearable devices and Cloud-based data and analytical systems designed to keep physicians and patients in control.

- **Remote cardiac care**: The remote cardiac care device provides a non-invasive remote care platform aimed at addressing the complex needs of arrhythmia monitoring, step down care and rehabilitation treatment for cardiac patients. The device consists of:

    ○ A light-weight ECG Device
    ○ A Device Aggregator that connects to multiple devices
    ○ A remote dashboard for monitoring the patient
    ○ An Administrative Console to regulate and control permissions and policies
    ○ 24X7 customer support

- **Remote care for expectant mothers**: The remote maternity care monitor keeps the baby connected with the doctor during pregnancy for mothers on the move and is especially recommended for those with common pregnancy complications and high-risk pregnancies. The monitor is similar to the cardiac care monitor in terms of system architecture but it monitors and records fetal heart rate, maternal heart rate, uterine contractions, blood pressure and blood glucose.

Both devices use proven signal processing algorithms and mobile technology to record and send continuous vital information about the patient (in one case the baby as well) to doctors in real time. This means the doctor can easily and quickly interpret the information (available on a dashboard over a mobile device), provide immediate attention and ensure the patient and baby are safe. The devices reduce the distance between patients and doctors, taking the stress out of hospital visits and the ever-present anxiety experienced by cardiac patients and expectant mothers.

A common problem with healthcare in emerging markets like India is the inaccessibility of clinics and hospitals. Not only are they statistically inadequate – India had 0.9 beds per 1,000 people versus a global average of 2.92 in 2005[1] – but, the care facilities deficiency reflects and leads other poor health care availability parameters. In rural India, clinics and hospitals can be located several miles from the patient. The distance is further accentuated by the lack of transport facilities and poor road infrastructure. In other words, visiting a hospital regularly is difficult, time consuming and inconvenient. In urban India the problem is quite the reverse. Health care facilities are more easily available, but urban congestion and long commute times makes them difficult to access.

However, mobile connectivity is excellent. This is reflected in the data for mobile connections. India had 616 million unique mobile users at the end of June 2016 with at least 1 billion mobile connections (SIM cards),[2] making India the second largest mobile market in the world. By and large, the Indian population can afford mobile devices and is comfortable using them – much more comfortable than with laptops, computers and tablets.

A mobile monitoring and management solution is easy to adopt, as there is an inherent dependency and trust in the mobile eco-system. Getting users to adopt a mobile solution for health care is simpler, requiring little persuasion and training. On the medical services side, a mobile monitoring device extends capacity and reach of the care giver.

The Wipro Assure Health™ research and development team acquired an intimate understanding of local conditions, patient circumstances, the accessibility of mobile devices and types, the propensity to use them and affordability. While these are excellent drivers for technical solutions, the key is user adoption. For this, Wipro looked at cultural markers such as the acceptance of mobile devices for other-than-communication needs and found that users were devising innovative means to extend the utility of their devices.[3] These were then used as inputs to design a device that was simple to use and could be kept and used in an unobtrusive manner,

---

[1] World Bank, http://data.worldbank.org/indicator/SH.MED.BEDS.ZS

[2] The Mobile Economy Report 2016, Global System Mobile Association (GSMA): https://www.gsmaintelligence.com/research/?file=134a1688cdaf49cfc73432e2f52b2dbe&download

[3] One example of user innovation is the widespread use of the "missed call". To understand this, think of the need to call your chauffeur with the car from the basement. The user gives the chauffeur a missed call – one ring could mean drive up to the office entrance, two could mean drive up to the office exit. The missed call costs nothing. The system has found widespread use for polling (a missed call to a number is recorded as agree/ disagree with a statement) and has other interesting applications in product feedback and service delivery.

> curbing social discomfort. Wipro's engineering expertise was used to ensure the device was robust and built for rugged Indian conditions. Through global team collaboration and attention to local customer needs, Wipro demonstrated market responsiveness and execution efficiency with a new range of mobile healthcare solutions.
>
> Authors: Balamurugan Kannan, General Manager of BFSI Global Delivery and Saksham Khandelwal, Member of Thought Leadership Charter, Wipro Limited, Doddakannelli, Sarjapur Road, Bangalore 560 035, India, November 2016.

Gathering Local Market Intelligence

Knowledge about international markets and customers is the key to unlocking ideas and creating valuable solutions. As shown in the Wipro case, the ability of an organization to tap into its international talent and market intelligence can make a great difference in local market success. The art of multicultural collaboration and knowledge-sharing enables geographically distributed teams to transform local market intelligence into strategic advantage. This requires a leader who can facilitate a global dialogue and knowledge-sharing in order to develop a successful international strategy.

While Front End Innovation and strategic planning are critical for idea creation and market validation, information-gathering and knowledge-sharing serve as the enablers throughout the project process. Since innovation is often inspired by customers and markets, one needs to look at the discovery process in order to share the experience with the market and to discover trends and practices which always come from physical experiences (Ikujiro and Konno 1998). The interactions that occur between HQ and subsidiaries during this process are instrumental to the local market experience with customers and the outcome for new products in international markets.

When gathering local market knowledge, there are formal and informal methods for organizations. The formal approach involves dedicated planning teams that are usually conducted on site through face-to-face meetings, in addition to survey tools sent to local teams to gather information regarding strategic planning and market needs. The project planning team gathers information from local sales teams while investigating local market opportunities. As noted by one of the study participants, "We have a kick-off meeting where we present the concept and have a list of features for local team members; we ask them to return with feedback or we may have individual or regional meetings with them to capture needs."

International travel and local immersions through team planning meetings and customer visits still remain the most effective for gathering local market intelligence. Project leaders can travel to the regions in order to hold team planning sessions where corporate management and local teams can review the plan and prioritize resources as well as budget allocation. It also provides the opportunity to meet with local customers and have more interaction with key accounts in international markets. Customer visits and dedicated forums provide a vehicle for sharing and testing new concepts. As shared by a senior global product planning manager: "We have good relationships with our international customers, we're successful in engaging them from concept to design to validation. There's a process around design validation and user/customer experience validation."

Discovery and exploration trips in key markets provide an effective learning experience and deep dive into local customers and practices. When I was working with a client on a consulting assignment, a discovery trip was organized for executives and cross-functional team members of the mobile communications group at a multinational electronics company based in South Korea. The group spent two weeks to gain a full view of the US market and customers through a series of visits, observations, mystery shopper excursions, ideation sessions, and in depth research on the dynamics of the US mobile market. The discovery experience provided valuable insights that were applied to a new model designed for local customers. It became a market success and pushed the company into the top five mobile brands in the US.

Informal communication methods are also helpful in gathering information from local team members. Project leaders and teams can rely on virtual communication methods such as conference or video conference calls, emails, or informal meetings conducted online or on site. The most effective communication tools and technologies include the use of video conference or telepresence, web conference and teleconference for virtual meetings. Spontaneous sharing through social media tools are also helpful for immediate updates during planning and execution activities.

A vice president of product marketing highlighted the opportunities generated by working more closely with local team members, "There are great marketing minds throughout the world where we need to leverage local team knowledge. If we look at how the local team tackles the challenge, then we can receive new results." The ability to listen to local markets and optimize local market intelligence within the team can generate new solutions and opportunities for the global innovation project.

Managing and Sustaining Team Performance

Under pressure to meet business performance objectives, cross-cultural teams need to meet expectations for increasing global and local sales revenue and market share, customer adoption and satisfaction, as well as product quality and performance. This means a continuous drive to manage cost and budget objectives, product and service quality, market positioning and differentiation, as well as timely delivery to market. In order to ensure team and project performance, leaders recognize the importance of effective cross-cultural interaction, communication, and collaboration skills. Yet few organizations incorporate global team collaboration and knowledge-sharing in project performance measures (Jensen 2014). Evaluation criteria for team performance are still driven by the project process and business results.

While there are fewer evaluations for team collaboration and knowledge-sharing, there are opportunities for individual evaluations on collaboration including peer and managerial reviews (Jensen 2014). As noted by a senior manager responsible for a global product line: "There are individual evaluations within and outside of the project. This is an important value at our organization." A global program manager noted there is a strict adherence to this behavior, explaining "there are key characteristics for measuring internal clients where we ask them about a person's collaboration approach…There are consequences for people who don't collaborate." Yet another manager explained that evaluations occurred within the team: "It's based on cross-team evaluations, so you provide feedback on the team members' performance, their strengths and where they need to develop."

Success metrics for the global innovation project are still primarily focused on traditional performance measures such as time to market, revenue and sales, quality of product, and customer base. A manager responsible for global product planning explained that "we are basically judged by profit, not process so much." However, there is a growing interest and need to move towards measures that focus on collaboration and communication for global team performance.

Introducing performance evaluations based upon collaboration competencies can support inclusive behaviors that enable multicultural team innovation. A global product manager emphasized the need for collaboration and knowledge-sharing evaluations within the organization, noting that "we see this of growing interest, it's important to the organizational culture. We also want to promote more community–building with internal recognition." As expressed by a senior manager responsible for global product planning: "Objectives play a more important role for promotions and appraisals. However, management believes competencies such as team motivation, influencing and inspiring, how you communicate your ideas, will be more relevant in the future – it defines who you are as a leader." Facilitating multicultural collaboration and innovation encompass global leadership skills and behaviors that will be crucial for orchestrating organizational performance and international market success.

Enabling a Team Climate for Local Innovation and Collaboration

The need to build in face-to-face time is essential for building trust and sharing knowledge that is critical to the global innovation project. The creation of a team space in live and virtual settings can facilitate collaboration and knowledge-sharing. Increased travel and interactions for initial meetings or kick-off events can stimulate creativity and exchange between team members. The project leader or a facilitator can design the process and ensure engagement in the early phases. There is also the opportunity to enjoy social networking and relationship-building through workshops and events during the kick-off meetings. A knowledge-sharing platform or system can sustain the dialogue throughout project collaboration with dedicated tools for creating, sharing, discussing, and storing knowledge. As shown in the case of an international product leader's experiences at Google, Twitter, and Airbnb, leaders can facilitate multicultural team innovation by optimizing international locations through creative exchanges, knowledge-sharing, story-telling, and multi-site project collaboration.

This chapter has presented the essentials of developing a global innovation climate that is focused on accelerating multicultural team performance and global launch success. Attention to international requirements during the five phases of the global innovation cycle, from creation to launch, enables the collaboration process throughout the FEI, NPD, and GTM stages. The international project collaboration process is shaped by the team climate values of entrepreneurial spirit, global team transparency, market responsiveness, and execution efficiency. Local customer connection and market intelligence are critical to the planning phase while marketing and sales readiness determine execution and launch performance. Leaders that pay attention to team and project performance through collaboration competencies and inclusive behaviors have the opportunity to strengthen multicultural team innovation.

> ### Creating a Global Innovation Climate at Google, Twitter, and Airbnb: Insights from an International Product Leader
>
> Every successful global company strives to create a climate for global innovation so they can expand their advantage in international markets. But, what exactly makes an innovative and collaborative team climate at hyper-growth companies like Google, Twitter or Airbnb? How do firms integrate multi-cultural teams while increasing organizational effectiveness? With past experience as an international product lead for fast growth multinational firms, I would like to share key practices from leading and developing global innovation initiatives in some of my previous roles at Google, Twitter, and Airbnb.
>
> During my work at **Google**, I led key international product initiatives including the company's localization efforts in 40+ languages and features for the Google home page. Today, Google has over 40,000 employees in more than 70 offices in 40+ countries. The company's mission statement sets the tone for their global vision: "To organize the world's information and to make it universally accessible and useful". Google managed to build a very strong internal brand while embracing international traditions from each of their offices and teams outside the U.S. – this becomes apparent when you visit a Google office in Gurgaon or Hyderabad in India in October where the festival lights of Diwali *and* Google-branded lava lamps greet visitors and set the tone.
>
> The company's commitment to their international teams and operations goes deeper than office decoration: Google uses simple yet highly effective web-based tools in clever ways to connect Googlers in all parts of the world making materials and data easily accessible for anyone who works at the company. It established micro-kitchens within each office to foster collaboration and ad-hoc meetings across teams. Google team members can share their professional profile with all employees where they share short descriptions of what they worked on and established in the past week. That simple but highly effective habit makes all company achievements and project work instantly accessible to any Googler in the world who may be looking to connect with someone who has been working on a similar project or talked to the same external partner.
>
> When leading international projects, new and creative ways to foster cross-team and international networking were made possible, for example a "random lunch" feature which connects local and visiting employees in any given location for a lunch meeting of up to six people. Google's 20% time policy encourages employees, in addition to their regular projects, to spend 20% of their time working on what they think will most benefit the company. In many cases these projects have been collaborations across international offices in multiple locations. From a greater business perspective, Google gives their country teams and local offices a great degree of liberty and budget authority, held together by their global OKR ("Objectives and Key Results") process which embeds every location into company-wide goal setting. Google's transparent goal setting process resulted in enhanced team and project performance on all levels of the

organization and particularly across geographies as joint objectives often connect teams in different parts of the world and encourage collaboration. In summary, I learned that full transparency and active exchange of individual and team goals, easy access to internal communication tools, and international company values were most valuable for leading cross-cultural teams and for developing a climate that encourages and enables international cross-team collaboration.

As an international project lead at Twitter, I was responsible for developing the vision, structure, and teams to support international product expansion. **Twitter** has 3,860+ employees in over 35 offices around the world which each reflect the regional and cultural spirit of their cities and countries. The company's inclusive, international mission is "To give everyone the power to create and share ideas and information instantly, without barriers.". More importantly, one of Twitter's ten core values is "Reach every person on the planet" to bake their international focus into their company DNA. To connect closely with team members in international locations, we could alternate times of weekly meetings and allow international offices to drive parts of the agenda broadcast live from their respective location. International user stories were shared frequently, allowing management teams at Headquarters in San Francisco to learn and understand local market practices. Easy browser-based access and video conferencing encouraged distributed teams to connect often and to collaborate closely.

Removing barriers and encouraging cross-team communication and collaboration made it easy for anyone in the organization, anywhere in the world, to connect in real-time. In some companies you may never talk to your colleagues in a remote part of the world. At companies like Twitter you are mostly not even aware where people are based. Geographic distances and time zones shrink when you are able to find a person you want to talk to, you see them online and you connect with them in an ad hoc video chat anytime within your browser, without planning or booking a conference room. My experience at Twitter demonstrated that a focus on the company's international user community and the easy access of internal communication technologies were most valuable for leading cross-cultural teams and for developing a globally inclusive team climate.

As Airbnb's Head of International Product, my role involved the development and expansion of the international group. **Airbnb** has nearly 3,000 employees in more than 19 offices all over the world. Their motto is "Belong anywhere". The company offers 2 million+ listings where it connects guests and hosts in 191+ countries and international revenue accounts for its largest source of income. Sharing user stories from guests and hosts across countries constantly informs product decisions and drives international team collaboration. Airbnb's mission and business has been inherently global from the earliest stages of company growth. International guests helped the company take a global perspective as they came from all parts of the world to stay in a few US cities and then became hosts in their countries of origin prior to the company's readiness for international expansion.

A number of international offices cater to specific needs of guests and hosts world-wide. Since the core market is mostly outside the US, country scouts and ambassadors were recruited in many places around the globe. They became important knowledge sources for Airbnb's product and sales teams since they represented local user needs and provided valuable insights for international requirements. Finding a lingua franca to define user processes is key for cross-team collaboration in building a service that works for customers in different countries. Airbnb defined their internal design language around user journeys which makes it easy to share host and guest experiences and collaborate across international offices.

Being global is Airbnb's business and teams have a strong collaboration focus amongst international offices. With meeting rooms that resemble international locations, team members in all offices are reminded of international markets and offices around the globe. The company issues travel credit as part of their benefits and encourages their employees to use the product and to experience it as host and guest every quarter. At Airbnb, I found that having company values on a global scale and constantly sharing user stories from around the globe has been most valuable for leading cross-cultural teams and for developing a team climate that actively encourages collaboration beyond office locations.

In my experience, companies are more likely to succeed in creating a climate for global innovation if they have a genuinely global mission that includes not only their international users in target markets but if they also have concrete and tangible internal values around collaboration with their distributed teams. Easy access and clever use of communication tools establish connection with international employees. Success factors include having international locations drive major parts of frequent company-wide employee meetings, sharing international user stories, encouraging cross-location projects and leadership with a mantra-like constant awareness of international teams. Ultimately, it helps to staff key executive and management positions with people who bring a solid professional profile and very diverse international background to ensure a globally-minded company DNA on all levels.

Author: Dr. Thomas Arend, Co-Founder and CEO, Savvy, San Francisco, California, USA.

Sources: Airbnb About Us page, https://www.airbnb.com/about/about-us, Accessed on December 1, 2016.

Company Locations page, Google, https://www.google.com/intl/en/about/company/facts/locations/. Accessed on December 2, 2016.

Company Overview page, Google, https://www.google.com/intl/en/about/company/, Accessed on December 2, 2017.

Company Overview, Twitter, https://about.twitter.com/company. Accessed on December 2, 2016.

Who We Are page, Google, https://www.google.com/intl/en/about/company/facts/. Accessed on December 2, 2016.

> **Knowledge-sharing Forums and Tools for a Team Innovation Climate**
> - Video cams to bridge distance between HQ and local offices. The cam can be set up in larger office areas where local team members meet or spend time. This allows a visual connection and regular opportunities to view team members.
> - Internal network or social media sites have helped cross-cultural teams get acquainted and build relationships throughout the project. They have also been used as ideation forums where users subscribe to each other's sites as well as using an indicator of interest or the number of likes in certain ideas. When top ideas are selected, the social network allows the identification and creation of teams to develop the concept.
> - Knowledge-sharing platforms are often project-driven with full access and ease of use for team members. Most of the software platforms are linked to the project process such as SharePoint, DocuShare, and Engage.
> - Knowledge networking platforms include options for identifying internal experts and collaborators for specific projects as well as early engagement through an idea incubator. The software platform is designed to help ideation and knowledge management for team members and stakeholders.
> - Video conferences and telepresence are popular tools for simulating live interaction, conducting team meetings, and promoting discussions.

Bibliography

Amabile, T.M. "Motivating Creativity in Organizations: On Doing What You Love and Loving What You Do", *California Management Review*, 40-1(1997): 39–58.

Amabile T.M. and Khaire, Mukti. "Creativity and the Role of the Leader", *Harvard Business Review*, Oct (2008): 101-109.

Internationalization vs. Localization, The W3C Internationalization Activity, https://www.w3.org/International/questions/qa i18n. Accessed on October 17, 2016.

Isaksen, Scott G. and Akkermans, Hans J. "Creative Climate: A Leadership Lever for Innovation", *Journal of Creative Behavior*, 45:3(2011): 161–187.

Isaksen, Scott G. and Lauer, Kenneth J. "The Climate for Innovation and Creativity in Teams", *Journal of Creativity and Innovation Management*, 11:1(2001): 74–86.

Jensen, Karina R. Global Innovation and Collaboration Study, 2014.

Nonaka, Ikujiro and Konno, Noburo. "The Concept of 'Ba': Building A Foundation for Knowledge Creation", *California Management Review*, 40-3 (1998): 40–54.

Part IV

Multicultural Innovation: Leading Change from Planning to Execution

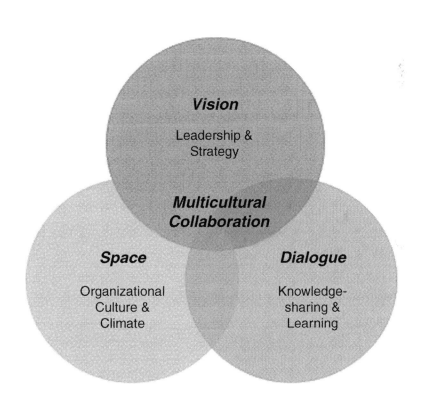

8

Facilitating and Orchestrating a Successful Transformation

"Our biggest issue is a lack of trust and communication," declared a senior director for global product management, "we need to build trust, build morale, and make them feel part of something important." Working for a leading US based multinational firm in the computer technologies and software sector, she felt her group and organization needed to increase communication. Meanwhile, the Vice-President of the group shook his head and couldn't understand the diverse requests from local markets, stating "the team in HQ does not always know how to evaluate input and link this to subsidiaries. Local markets also propose initiatives without justification, where we need to substantiate and provide business reasons."

Across the Pacific, regional and local management leaders in Asia were very frustrated concerning the current process for global product innovation. A regional leader traveling between China and Japan stated "the organization is disconnected with too many silos since HQ has a US-centric focus where the corporate team is a closed circle and does not interact with the local teams," he frowned and then added "there is no resource commitment and the HQ management team is not supportive." There were serious problems with cross-cultural understanding and learning where there was no collaboration in the front end, especially for ideation and strategic planning. As a result, there were product quality and application issues with dissatisfied customers, lower market demand, and poor sales results in the Asia region.

"I would like to be regarded as an expert for my local market and customers, so I can provide value and solutions," a local team manager in Vietnam explained "if something is promised, HQ needs to ensure team members can help deliver a solution so that it is valid and we can build trust with the local team and customers." He found the challenges of time, speed to market, and a lack of organizational transparency had a negative impact on global team collaboration. However, his team had a great interest in increased knowledge-sharing and collaboration with the management team in HQ in order to ensure local market success.

Back in US HQ, the global vice-president responsible for products was dealing with additional criticism from local teams in European subsidiaries since there was limited involvement in the concept creation and strategic planning phases. The HQ management team and local teams in Europe could not reach an agreement on the new concept features and specifications. As a result, the product development and launch phases were delayed until an agreement could finally be reached through mediation.

In view of the challenges facing team collaboration across geographies, the vice president acknowledged that "we need to articulate the use case, understand better what is needed and understand the local context". There would need to be more focus on market insights, need articulation, and cross-cultural learning in the regions in order to understand local customer preferences. Moving forward, there were plans to ensure global integration of local teams in the front end and to provide more marketing and sales support. Looking ahead, the vice president noted that "we need to make decisions on where to invest, globally or regionally, and ensure the regions have more autonomy and participation."

Moving from Global Reach to Global Readiness

Are you ready to change the world with your great innovation? The answer is largely determined by the readiness and capabilities of leadership, teams, and the organization. There are the pressures to stay globally competitive and understand international market needs while navigating a multicultural and digitally connected business environment. Customers in every geographic location expect timely and localized products, services, and solutions. The question then becomes how can you collaborate effectively across cultures in order to ensure international market success?

Before you plan your world tour, make sure to take inventory of your organization's global readiness capabilities. Determine who will be onboard for your journey, including the core team with members from functional groups as well as the regional and local team managers from key markets.

Amongst the functional groups, determine readiness for international support and cultural understanding across the value chain, including talent needed for R&D, design and development, production, marketing, sales, operations, and support. Then identify the technology infrastructure and support needed for driving a global knowledge platform, communication process, and tools that sustain project collaboration. When working with clients, there are three key questions that help me determine if they are ready to launch:

1. **Do you have a global plan and process that is shared by teams worldwide?**
2. **Are teams and regions aligned across functions and cultures?**
3. **Are your teams ready to communicate and execute in every geographic location?**

These three questions allow the identification of internal needs for ensuring global innovation readiness. It will be critical to evaluate and develop the necessary organizational resources and capabilities that allow the successful creation and launch of new concepts to international markets. It also goes hand in hand with the global market assessment when identifying suitable products and services as well as primary and secondary markets. Next step is the development of the global plan in terms of (1) Business Feasibility – defining launch objectives, target markets, the revenue plan, and securing local budget allocation and (2) Market Opportunity – understanding local customer, market, and concept needs.

Orchestrating Global Innovation Through Organizational Mechanisms

Leading global innovation requires an open mind and collaborative focus in order to navigate changing landscapes and organizational dynamics. An international and digitally connected environment requires that leaders have the capabilities to facilitate multicultural collaboration as well as orchestrate the global innovation cycle, from concept creation to launch. On one hand, there are the leadership behaviors and skills required to facilitate knowledge-sharing and learning through each phase of the global innovation cycle. On the other hand, leaders will need to orchestrate organizational mechanisms and project routines that support and strengthen global

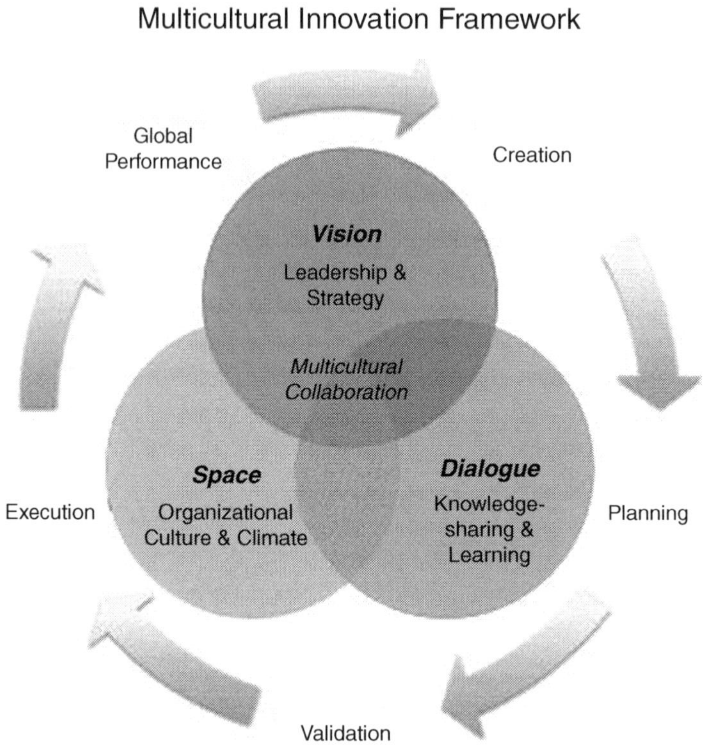

Multicultural Innovation Framework

Fig. 8.1 Multicultural innovation framework in action
Source: Dr. Karina R. Jensen, Global Minds Network, 2017

innovation performance. As noted in Fig. 8.1 and Chapter 1, The Multicultural Innovation Framework serves as a valuable reference and guide for your journey in leading global innovation.

The framework shows the multicultural collaboration drivers – Vision, Dialogue, and Space – that link to the key phases in the global innovation cycle. This book has provided detailed insights on the mechanisms that are crucial to achieving organizational performance and global market success. In order to highlight the critical success factors, an overview of key organizational mechanisms is presented along with a review of the collaboration drivers for multicultural innovation. Successful performance is determined through the interdependent orchestration of organizational mechanisms:

- **The innovation strategy and its focus on global and local participation.** When developing the global innovation strategy for introducing new products to international markets, leaders need to engage local teams as planning partners in order to allow for strategic co-creation and collaborative planning. A top-down global strategy with minimal participation by local and regional teams often results in a disconnect with local markets and a negative impact on international sales.

- **A knowledge-sharing structure and roles for local teams.**
A formal knowledge-sharing structure should be aligned with the project process through clear roles for local team members within the global innovation project cycle phases – from planning to ideation to validation to execution. Organizations that understand how to transcend organizational structures and layers in order to optimize the global network can accelerate knowledge-sharing and learning around the world.

- **Communication vehicles that integrate face-to-face and online interactions.**
When managing the global innovation project, a dedicated communication system, process and tools should be an integral part of project collaboration. Consider opportunities for face-to-face interactions through regular meetings supported by a knowledge platform with communication technologies for consistent communication throughout the project.

- **The organizational culture and its innovation values.**
An organizational environment that lacks cultural understanding and global transparency will find a disconnect between HQ and subsidiaries worldwide. In order to create a stronger foundation for project collaboration, the organizational culture serves an important role in developing values for cultural empathy and appreciation for diversity, collaboration through team transparency across regions, and creativity through cross-cultural learning and ideation.

- **The team climate for multicultural innovation.**
In supporting team collaboration throughout the global project, the innovation climate needs to support the values of market responsiveness, global team transparency, entrepreneurial initiative, and execution efficiency. A climate that lacks transparency, responsiveness, and initiative will have a negative impact on execution and performance.

When you've identified these mechanisms within the organization, it's time to evaluate and determine their application to the global innovation project. You need to identify which mechanisms will be supportive of innovation and which mechanisms may block project collaboration around the world. The purpose is to optimize mechanisms that support global innovation and collaboration while addressing weaker mechanisms that may challenge the process.

Optimizing and Facilitating Multicultural Team Collaboration

The change levers for facilitating global innovation and collaboration are presented in the Multicultural Innovation Framework as shown in **Chapter 1** and Part IV: (1) Vision through leadership and strategic co-creation; (2) Dialogue through knowledge-sharing and learning; and (3) Space through organizational culture and team climate. These change levers are closely aligned with the global innovation cycle, from creation to execution and global team performance. Leaders and managers need to facilitate knowledge-sharing and learning in the front end of innovation (creation, planning, and validation phases) while orchestrating the project process from planning to the execution phase.

- **VISION – Lead Through Strategic Co-creation**
  Inclusive Leadership for Facilitating Knowledge-Sharing
  Leadership is essential in communicating the vision and providing the road map for multicultural and geographically distributed teams. Common goals and objectives need to be set by the project leader and the team in order to create a coherent vision and understanding of the expected outcome. Global project leaders require openness to diverse cultures and perspectives. There is a greater need for trust-building and frequent communication throughout key project phases, from ideation to execution. The project leader needs to ensure the time and space for team members to share knowledge and exchange ideas.

    The opportunity to learn about local practices and perspectives requires face-to-face interactions with an emphasis on relationship-building. In order to engage in Front End Innovation, an open and safe environment with a collaboration focus is important for the team. Team members need to build trust in order to share, discuss, and resolve problems. It's important to voice needs, address time delays, issues, and communicate team status. In managing time zones, social media and virtual meetings can address regular communication in real time. Cross-cultural understanding and team cohesion is improved by offering learning and knowledge-sharing opportunities in a consistent manner.

    Leading global innovation requires special attention to behaviors and skills that facilitate multicultural team collaboration throughout the project process. Special attention needs to be placed on the new leadership behaviors required for each phase of the global innovation cycle, from concept to market. Developing practices for empowerment, inclusion, direction, and communication facilitate multicultural collaboration for creation, planning, validation, and execution phases. The ability to facilitate knowledge-sharing while orchestrating the global innovation process creates multicultural team performance.

Strategic Co-creation Through Engagement in Front End

Introducing opportunities for strategic co-creation can improve team engagement and collaboration. It's important to emphasize collaborative and entrepreneurial roles for local team members during the creation, planning, and validation phases of front-end innovation. This increases knowledge-sharing and collaboration which positively impacts project performance for improved time to market, product localization, customer demand, and local sales. The opportunity to engage team members in collaborative and entrepreneurial roles during front-end innovation creates increased trust and motivation.

The strategic planning phase is pivotal when engaging local teams as planning partners in order to create a shared understanding of strategy and implementation. Team interactions facilitate the sharing of local market knowledge, cultural understanding, and the creation of new ideas that lead to international market solutions. It enhances local customer connections and the capture of critical knowledge for market responsiveness. Creating new market opportunities requires more focus on co-creation at the local level with integration at the global level.

- **DIALOGUE > Create Interactive Conversations**
Local and Global Knowledge-Sharing Through Active Listening
The project collaboration process requires systems and tools that allow opportunities for exchange and dialogue. Consistent communication can be achieved through regular meeting opportunities throughout project collaboration. Global project leads as well as regional and local managers can facilitate communication between HQ and subsidiaries. In order to improve understanding of local requirements, clear ownership with roles and responsibilities concerning global and local needs can be effective in capturing commitment for the project.

When facilitating Front End Innovation activities with cross-cultural team members, special attention is needed for the creation and planning phases due to the cultural implications of knowledge-sharing. These elements include the structure and communication context, the role of power for the leader and the team, the degree of openness to share ideas, the opportunity for initiative-taking, and the response and feedback process. Allowing time for reflection and concept creation can facilitate knowledge-sharing and collaboration with teams worldwide.

Cross-Cultural Learning Through Integrated Communications

Improved cross-cultural understanding can be achieved through team engagement and dialogue concerning cultural learning challenges and opportunities. There are unique opportunities to learn from diverse cultural perspectives in creating concepts and a project process that optimize the global innovation cycle. This requires greater emphasis on trust-building, team participation, interactive dialogue, and a shared understanding of the global strategy. Building trust and motivation across cultures and geographies demands a focus on the ability to actively listen and respond. Leaders who can demonstrate active listening and responsiveness will more effectively build trust and motivation through cultural and market understanding.

Face-to-face dialogue through country visits and exchanges create powerful opportunities for building trust and motivation. Developing relationships and sustaining an inclusive dialogue requires interpersonal team interactions. It's also a necessity for validation of new concepts in collaborating closely with local teams and customers. Being immersed in the local environment enhances cross-cultural learning and understanding of market needs and opportunities. Visual communication and collaboration technologies create a blended learning environment for sustaining dialogue around the world.

- **SPACE > Develop an Environment For Project Collaboration and Communication**

A Work Environment That Supports a Global Innovation Culture

The creation of a work space and environment conducive to collaboration requires both a global innovation culture and team climate. A global innovation culture provides the foundation of common values while the innovation climate enables the network and connection to international customers and markets. Nurturing innovation in international organizations requires the development of three values: cultural empathy, collaboration and creativity. Leaders, teams, and organizations are creating cultural empathy through an emphasis on culturally diverse talent and global teamwork. Creativity demands idea generation and innovative thinking by optimizing cultural knowledge while collaboration requires transparency and knowledge-sharing for building an inclusive work environment.

The global innovation culture is shaped by organizational routines that support these three values. Organizations and leaders can create an inclusive mindset with international mobility and face-to-face interactions through regular travel and international forums. In addition to shared work spaces, an effective project process and communication tools can enhance multicultural collaboration. The creation of both live and online spaces facilitates knowledge-sharing during the global innovation project. Increased travel and interactions focus on the initial meeting or kick-off event where a live work space can stimulate creativity and exchange between team members.

The Cultivation of a Team Innovation Climate
Creating the optimal team environment for nurturing an innovation climate requires four values that support an entrepreneurial spirit, global team transparency, market responsiveness, and execution efficiency. Entrepreneurial spirit allows for initiative and new ideas while market responsiveness provides a focus on local customer and market needs. Global team transparency enables knowledge-sharing and collaboration while execution efficiency drives performance.

A project process and communication system that support the global innovation cycle are essential to team performance. The FEI, NPD, and GTM stages should be ready for international and multicultural environments. Global innovation projects demand attention to strategic planning and local co-creation, internationalization and localization of the new concept as well as marketing content, and sales readiness worldwide.

Team climate is enhanced through work spaces that enable collaboration through tools for visual expression, ideation, and knowledge-sharing. The leader as knowledge facilitator can enhance the process and ensure engagement from creation to execution. There is the opportunity to enjoy social networking and relationship-building through workshops and events during kick-off meetings. Knowledge platforms that incorporate dedicated tools for communication and collaboration such as web conferences, video conferences and telepresence can sustain dialogue around the world. Using communication technologies and tools to enhance regional and cultural interactions through a blended learning environment may be the best platform for promoting multicultural collaboration.

- **Global Team performance**
  When developing evaluations for global teams, practices and skills that incorporate collaboration and knowledge-sharing should be considered as measures for project performance. Performance evaluations need to place a greater emphasis on multicultural intelligence, collaboration and knowledge-sharing competencies in order to support a global innovation culture. Success metrics require a balance of business performance and team performance measures.

The ability of global leaders and teams to facilitate multicultural collaboration during the front-end of innovation can accelerate global innovation performance for international markets. Paying attention to the key drivers of Vision, Dialogue and Space helps leaders effectively orchestrate the global innovation cycle. Figure 8.2 presents a summary of key practices for leading global innovation through multicultural collaboration. Establishing a common vision, values, and language provides a foundation for the leader to act as knowledge facilitator in engaging geographically distributed teams in the front end. The ability of leaders to listen and respond creates dynamic and interactive conversations through live and online team interactions. Developing a global space for team communication sustains collaboration and transparency throughout project collaboration

Multicultural collaboration practices are most effective when there is an interdependent facilitation of vision and dialogue as well as orchestration of the space needed for project collaboration (see Fig. 8.2). A shared vision through inclusive leadership and strategic co-creation empower teams worldwide. An open and interactive dialogue develops trust and cultural understanding in order to improve connections with teams and customers in local markets. A safe and transparent creative space for team collaboration enables the transformation to a global innovation culture through cultural empathy, creativity, and collaboration. This leads to a team

Fig. 8.2 Multicultural collaboration practices
Source: Dr. Karina R. Jensen, 2017

innovation climate that inspires entrepreneurial initiatives and team transparency for increased market responsiveness and execution efficiency around the world.

Connecting Global and Local Knowledge, from Concept to Market

When creating, and launching new concepts to international markets, organizational communication flow moves primarily from global headquarters to subsidiaries with variable levels of feedback. There is less communication initiated by teams based in local subsidiaries and markets, and minimal communication between local subsidiaries (Jensen 2014). In order for leaders to succeed in engaging local team members in planning and execution phases, the global network needs to be optimized to ensure a balanced and dynamic communication flow between management teams in headquarters and local teams in subsidiaries worldwide. Challenges are compounded by time pressure, a growing demand for local support and resources along with complex organizational structures and layers across functions and geographies. The challenge for leaders and teams is to create and sustain dialogue that taps into local knowledge within a global and dynamic network.

In the age of the collaborative economy and a profusion of communication technologies, it should be easier to sustain dialogue across countries. However, geographic distances have also made online communication challenging due to the need for translation and interpretation of cultural communication styles. As noted in previous chapters, digital communication and social media tools can improve global collaboration and transparency. However, there is still the need for face-to-face communication and on site team collaboration in order to build trust and cultural understanding. Leaders and teams will need to consider the development of a project communication structure and process that support multicultural collaboration through an effective blend of onsite and online communications.

When designing a global communication architecture, there are three levers that empower multicultural collaboration: social networking, knowledge-sharing, and cross-cultural learning. As shown in Fig. 8.3, the alignment of these communication activities with the project collaboration process can strengthen global team performance while optimizing the internal knowledge platform and network worldwide.

Social Networking

When collaborating around the world, the art of social networking is not solely based on online interactions as demonstrated by social media platforms. Leaders responsible for global projects and teams find that face-to-face communications are more effective for building trust and relations (Jensen 2014). There is a need for balance with an equal amount of time devoted to formal, informal, and online communication. In the early phases of creation and strategic planning, it's especially important to connect all of the team members worldwide. Social networking is especially valuable in connecting cross-cultural and cross-functional groups in understanding roles and knowledge-sharing opportunities.

As discussed in Chapter 4, the cultural context has great influence over the amount of social networking that should be devoted to global team communication. High context cultures such as Brazil, Italy, Japan and Russia require more socialization and relationship-building. On the other hand, low context cultures such as Germany, the Netherlands, Scandinavia, and the US require less relationship-building and more focus on project collaboration due to the emphasis on business objectives. Thus, you will need to consider cultural barriers in terms of context as well as structural barriers in terms of the organizational hierarchy and work environment. There will always be the challenge of time for networking due to fast-moving innovation projects where social activities need to be built into the project collaboration process.

Formal

The formal mode of social networking is often demonstrated through specific business and project meetings, forums, and video conferences. Depending on the organizational culture, there may be a greater emphasis on formal meetings to gather teams and provide a venue for discussion of key project topics. There's also the need for recording key points and results as well as actions for team meetings that involve geographically distributed teams. Due to culture, distance, and language differences, precise agendas with follow-up notes and actions help clarify understanding across regions. Formal meetings are also valuable for reaching a common understanding for initial concept presentations, strategic planning, and key issues related to project execution in local markets.

Informal

The informal mode of social networking is defined by spontaneous encounters and interactions. The opportunity to create a work environment with collaboration spaces, lounges, or cafés provide team members with the opportunity to meet face-to-face and enjoy brief conversations. From the water cooler moment to the café update, informal conversations add value to trust-building and relationship-building. Most ideas or collaboration opportunities are often inspired from informal meetings. Cross-cultural team members may not feel comfortable expressing all of their feelings or perspectives in formal meetings due to time, language, and sensitivity to innovation roles. Social moments and informal conversations can therefore facilitate communication in a more open and casual environment. International meetings can especially benefit from opportunities to socialize, whether it's networking and connecting with colleagues or celebrating project milestones. Informal networking provides opportunities to build trust and to strengthen relationships.

Virtual

Although social media platforms and networks are in abundance and used frequently for personal and customer communication, their popularity does not reflect actual practice within multinational organizations (Jensen 2014). Attempts to increase social networking through technology platforms and online networks have been met with mixed success due to cultural and structural barriers. Since on site interpersonal communication is still preferred, social networking online is more effective when screen interactions can occur through video conference, telepresence, and web conference.

Social media platforms that succeed in engaging team members worldwide can reduce time in capturing data and information while providing more resources for problem-solving and finding solutions. However, there is not yet a completely integrated platform that ensures social media efficiency and multicultural team engagement. Social media tools may sometimes serve as a better solution if technology platforms are not universally accepted or used. An internal platform that offers key communication technologies for project collaboration can enable the team to connect and share ideas. This includes real time access online to enable networking and responsiveness.

Social media tools that allow more expression through spontaneous chats and discussions can enhance social networking. Organizations are exploring ways to optimize mobile devices through immediate connections to social networks, forums, collaboration and cloud tools. In addition, visual communication tools are very popular for networking across cultures, such as video sharing, internal YouTube platforms where teams can post videos or share through blogs, social media platforms and message texting. There is also the use of chat and discussion tools as well as wiki tools and databases for searching relevant information from team members in the organizational network in order to exchange information and stimulate conversations. In creating a cohesive social networking system, there is more structure needed to share information and provide online exchanges.

Global and Local Knowledge-Sharing

International organizations demand increased knowledge-sharing in order to learn and connect with local customer needs. Leaders find there is not sufficient sharing of knowledge inside their organizations. There are challenges of a knowledge disconnect between HQ and subsidiaries where cross-cultural and cross-functional teams do not have full access or transparency (Jensen 2014). There's a greater need for local co-creation and validation of new concepts where go-to-market pressures and an execution focus may block opportunities to fully engage with local teams and customers. Since knowledge resides within team members, leaders have an important role in facilitating a collaborative dialogue where team members around the world can make connections, share and learn.

A fast-paced and global business environment makes sharing relevant knowledge for multiple markets complex and challenging. There needs to be sufficient knowledge for understanding the opportunity and context while avoiding an overload of information. An effective infrastructure with a shared knowledge platform can promote pro-active sharing rather than reactive feedback. A structured knowledge-sharing process with key mechanisms allows for interactive communication flows and informal sharing throughout international project collaboration, from ideation to execution and lessons learned. Using a common format with concise messages can facilitate reception and understanding across cultures.

The phases of the global innovation cycle require an integration of face-to-face and virtual communications. Real-time communication is essential whether in person or online in order to promote team interactions. Open and transparent communication can be sustained through day-to-day interactions, live events, presentations, regular meetings and collaboration sessions. A collaboration platform should be designed for both a social

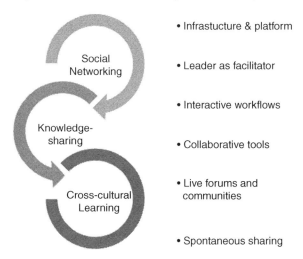

Fig. 8.3 Connecting global and local knowledge

architecture and a technology architecture in order to enable sharing through effective collaboration tools that are suitable for multicultural and multilingual teams. This platform should serve as a central knowledge source and window to past, present, and future activities.

A high-performance knowledge-sharing platform should be designed for multicultural collaboration throughout the global innovation cycle. It needs to take into account a multicultural and multilingual workforce located throughout the world. It's important for leaders and teams at global and local levels to understand and appreciate the value of sharing knowledge to enable collaboration. Engaging with local teams results in valuable insights on customers and market needs. Global and local knowledge-sharing is therefore an integral part of an innovation culture and organizational performance (Fig. 8.3).

Cross-Cultural Learning

There is great interest for cross-cultural learning in order to effectively collaborate with international teams and customers yet a majority of leaders find their organizations are not providing sufficient resources and support (Jensen 2014). Cross-cultural learning opportunities require formal and informal learning as well as online tools that can increase cultural awareness and understanding. Organizations that are not able to provide cultural training often fail to understand the impact on organizational performance and international market results. The consequences are lack of time, budget allocation, resources, and mobility for teams to benefit from global training programs.

Organizations that are succeeding in creating a global innovation culture and team climate have paid attention to increased mobility and exchanges between regions and countries. They have realized the importance of cultural knowledge prior to front end innovation in order to avoid cultural issues and concepts that do not meet local customer needs and international market opportunities. As a result, there is more emphasis on culturally diverse teams and integration between HQ and subsidiary locations. This requires sufficient support and resources with activities that enhance project collaboration throughout the global innovation cycle. Cross-cultural learning programs provide teams with exposure to different cultural views, facilitate knowledge exchange, and enhance communications and the collaboration process. Cross-functional teams can benefit from training for improved understanding of regions and cultures, especially local practices, applications, and technology options that result in successful market solutions.

Cross-cultural learning shapes the attitudes and openness of leaders and teams to the needs and interests of international markets and customers. Organizations that succeed in developing and implementing comprehensive education programs can accelerate knowledge of local markets and customers that are critical to business decisions. In developing a company-wide program for cross-cultural learning, there should be attention to three primary areas – formal learning through education and training, informal learning through team collaboration, and online learning through information and interactions delivered through social media and learning platforms.

Formal

Formal learning is developed through internal programs and events that serve as complementary training to leaders and teams participating in global innovation projects. They vary from learning forums and sharing of best practices to education from team members on their cultures and local market practices. An increased emphasis is being placed on cultural immersion programs such as focused learning tours to key markets and exchanges through job rotations in various countries. There is also the opportunity to integrate leaders and teams through orientation and onboarding training in cultural behaviors and practices. These programs allow leaders and teams to spend more time in specific countries and key regions in order to improve cultural understanding.

Formal education and training programs should take into consideration the Global Leadership Development Journey layers presented in Chapter 2. First and foremost, core leadership and team values need to be addressed due to the importance of self-knowledge and the cultural self-reference criterion (how one projects his or her cultural values onto others). The second layer offers the opportunity to learn about national cultures and practices through country profiles as well as cultural differences through cultural dimension models. Then there is the role of cultural interactions and how to develop leadership mindsets and cognitive behaviors that facilitate cross-cultural team communication. Finally, there is the context layer in leading and facilitating multicultural collaboration and innovation.

A structured learning program that considers these layers should also integrate the three modes of formal, informal and online learning. Organizations will need to learn how to develop competencies for working in a cross-cultural and global environment. Understanding cultural practices, context and language will enable leaders to more effectively collaborate throughout the global innovation cycle, from creation to execution. Internal knowledge can be optimized through workshops with international managers and cultural experts, in addition to partnerships with learning institutions, universities and business schools. In order to ensure integration and learning between teams in HQ and subsidiaries, organizations will need to design programs that incorporate international mobility, country visits, job sharing and exchanges, resources, and tools.

Informal

Informal learning often becomes part of the global innovation project process through face-to-face meetings and team-building practices. Regional and local meetings and events that are tied to innovation projects allow cross-fertilization of ideas and knowledge from team members worldwide. Throughout project collaboration, there is the opportunity to listen and learn from team members through informal meetings, briefings, and exchanges as well as onsite and online communications. Through international travel and exchanges, team members worldwide can interact at HQ or subsidiary locations by sharing best practices and success stories on a regular basis. Cross-cultural learning opportunities can be built into the project collaboration process through spontaneous updates at weekly or monthly meetings with the regions. Through cultural immersions and country visits, there is also the opportunity to explore local markets during day-to-day interactions and experiences.

Online

There's a broad offering of online learning tools and assessments available for global team learning and collaboration. Cross-cultural training firms and educational institutions offer their own instruments such as web portals with cultural profiles and online assessments for global leadership and team collaboration. In addition to learning from cultural websites and portals, organizations are optimizing internal knowledge platforms for developing repositories that contain country profiles and cultural practices. Social media tools and communication technologies are creating connections with local teams where informal learning can occur through country experts. Online learning systems and MOOCs provide additional opportunities for leaders and teams to tap into cross-cultural training through courses, webinars, and video clips. In order to create an optimal cross-cultural learning program, it is best to design custom training that considers the Global Leadership Journey phases and the integration of formal, informal and online learning activities.

Facilitating and Orchestrating Multicultural Innovation

Leaders and teams are the catalysts for driving change and enabling global innovation through multicultural collaboration. As shown in this chapter, an increasingly critical role for leaders is the ability to serve as facilitators of knowledge-sharing in order to promote social networking and cross-cultural learning. In creating a vision for multicultural innovation, leaders have the opportunity to engage teams and optimize knowledge for strategic co-creation worldwide. When facilitating an inclusive dialogue in the Front End Innovation stage, team leaders can nurture a global innovation culture and climate that creates the optimal space for project collaboration. A multicultural innovation mindset is required by leaders in developing new behaviors and competencies that meet the demands of a global innovation cycle and international markets.

In an increasingly networked and collaborative business environment, there is a need for inclusive leadership that inspires new ideas and facilitates knowledge-sharing among diverse groups. Leadership is more about inspiring collaboration through behaviors that are empowering, inclusive, communicative, and directive. Leaders will need to master the art of facilitation in order to promote open dialogue and knowledge-sharing in a complex multicultural organization. Inspire teams through vision, dialogue, and space. Co-create the vision and facilitate the dialogue and develop the space necessary for ideas that optimize the knowledge and talent of your cross-cultural and cross-functional teams.

While global leaders have the opportunity to facilitate front end innovation, they also face the demand for effective execution and innovation performance in international markets. As shown in the Cisco case, it is a delicate balance of the leader as facilitator of knowledge-sharing and orchestrator of organizational innovation. Effective orchestration of the innovation process from concept to market requires attention to organizational mechanisms and project routines. Leaders will need to ensure organizational readiness through the key levers discussed in this chapter: strategic planning, knowledge structure, communication process, organizational culture and team climate. The art of facilitation and mastery of orchestration will shape leadership competencies and organizational capabilities for multicultural innovation and collaboration.

Transforming How People Connect and Collaborate at Cisco

Founded in San Francisco in 1984, Cisco has enjoyed international growth opportunities as a global market leader in networking equipment. The company has shaped the future of the Internet with products serving the increased need to manage bandwidth and internet traffic. Transforming how people connect, communicate, and collaborate, Cisco has nearly 74,000 employees with more than 200 offices worldwide.

With an organizational culture based on empowerment, engagement and innovation, the company regards employees as part of a global community that values inclusion and diversity in a connected workplace. It has an impressive Corporate Social Responsibility program that aims at developing an inclusive digital revolution where everyone can access technologies and create solutions regardless of their location. This initiative intends to impact one million people in the next decade by harnessing the internet and enabling teams to serve as global problem-solvers.

Cisco has an innovation climate that is focused on market responsiveness and execution efficiency. Front end innovation activities are primarily centralized with a focus on both radical and incremental innovation. The company strategy is to lead customers in their digital transition through secure, automated, and intelligent solutions that strengthen digital connections. Cisco offers an integrated innovation approach in building its own solutions, acquiring new companies, partnering with technology and services partners, investing in promising start-ups, and co-developing with customers and innovators. This global effort includes nine innovation centers, a $2B investment portfolio, and 150 global start-up investments and funds.

The challenge for global product teams is to ensure quick execution and effective localization for key geographic markets around the world. In the past, the networking company used a global strategy where concepts were primarily conceptualized and designed at headquarters with directives from the HQ global product leadership teams to the theatres (subsidiary offices) and local teams for execution. Communication primarily relied upon web technologies without face-to-face meetings which further reduced opportunities for relationship-building and local market understanding. As a result, global concepts were not always adapted to local market needs and customer preferences. There was also the risk of delays for product introductions as well as reduced market and sales opportunities.

Today there is more attention to engagement of local teams in the front end with emphasis on strategic planning and co-creation. A global planning framework is used for identifying requirements when meeting with local teams, sharing and exploring the product concept and local market opportunities. Increased travel and joint planning sessions align global business objectives with local market opportunities, including reviews and prioritization of resource and budget allocation. In addition to web meetings via Cisco WebEx, an internal knowledge platform combined with telepresence for international meetings allow teams to optimize Cisco technologies for improved collaboration.

Global product leaders and management teams in HQ have developed a co-creation strategy where joint planning takes place with local teams at the subsidiary level. Organizational culture values of collaboration and engagement are applied in creating a more open and transparent environment. Through more participatory planning, local team members are enjoying a collaborative role where they can shape ideation, planning, and validation for global concepts. This approach creates inclusive leadership which results in increased motivation and engagement from local teams. Leading the way for global innovation, Cisco has demonstrated high performance and effective execution in time to market, localized solutions and continued growth in mature and emerging markets.

Sources: Cisco Annual Report. http://www.cisco.com/c/en/us/about/annual-reports.html. Accessed on December 9, 2016.

Cisco Corporate Overview. Used with the permission of http://thenetwork.cisco.com/. Accessed on December 9, 2016.

Jensen, Karina R. Global Innovation and Collaboration Study, 2014.

Jensen, Karina R. Local Innovation and Collaboration Study – Asia, 2015.

Global Readiness Audit

Vision
1. Have you developed an inclusive vision and plan?
2. Have you optimized ideation and knowledge-sharing in local and regional markets?
3. Does your project process ensure effective local market execution?

Dialogue
4. Do you have an internal communication process to engage and align teams worldwide?
5. Do you have a consistent collaboration framework and process worldwide?
6. Does your global communication plan connect teams to local customers?

Space
7. Do you have the platform and tools for global ideation and knowledge-sharing?
8. Is your team knowledgeable and responsive to international innovation needs?
9. Does your team have the knowledge and skills necessary to collaborate across cultures?
10. Do you have a global readiness program that provides you with a complete solution?

Bibliography

Jensen, Karina R. Global Innovation and Collaboration Study, 2014.
Jensen, Karina R. Local Innovation and Collaboration Study – Asia, 2015.

9

The Future of Multicultural Innovation and Collaboration

The global business environment is in a turbulent state with great implications for leaders, teams, and organizations. Globalization and international trade have opened borders and created opportunities for collaboration across nations while driving worldwide growth through the last half century. However, for the last, few years there has been a slow yet steady backlash to globalization due to outsourcing, foreign direct investment, and shifting import/export balances. There are concerns about increased unemployment, economic and environmental disadvantages for host countries, immigration and integration of a culturally diverse talent pool.

Global innovation has been criticized due to the commercialization and proliferation of products and services that do not benefit all countries, an increasingly competitive business environment, and gaps in innovation competencies and capabilities between mature, emerging, and developing economies. There is a multipolar world of research and innovation where the majority of activities are still concentrated in high-income economies as well as select middle-income economies such as Brazil, China, India, and South Africa (Global Innovation Index 2016). Other middle income as well as low income economies need to accelerate innovation through education, government and business investment, and entrepreneurial opportunities. Multinational corporations need to move from centralized innovation at headquarters and home countries toward decentralized innovation that optimizes the global value chain across geographies.

The challenges and questions faced by the world provide unique opportunities for leading multicultural innovation and collaboration. Instead of turning away from an interdependent global community and solely focusing on national interests, leaders and their organizations need to consider opportunities to share ideas, knowledge and talent that can benefit countries, regions and the entire planet. Rather than focus on unlimited commercialization and exploitation of products, services, and technologies that do not add value to international markets, there is the opportunity for organizations to solve critical challenges and problems by delivering solutions with increased value to local markets and countries in which organizations operate.

We live in a time where leaders and organizations can serve as powerful catalysts for change in creating a vibrant global community where culturally diverse talent is celebrated and optimized rather than ignored and feared. Living in an increasingly interconnected and interdependent world provides the opportunity to learn and benefit from a wealth of diverse perspectives and practices. Yet this potential is not fully optimized within our communities and organizations. Culturally diverse views bring valuable insights for creating local and global solutions. In a constantly changing business environment, leaders need to listen to local voices in order to facilitate and orchestrate innovation around the world.

In order to develop and prepare leaders for the demands of facilitating and orchestrating global innovation, there needs to be greater emphasis on leadership and team development through social networking, knowledge-sharing, and cross-cultural learning. There is a growing demand for competencies in multicultural collaboration and innovation.

Global leadership skills that are noted as most critical yet leaders considered themselves least effective include (1) Leading across countries and cultures, (2) intercultural communication within international business environments, (3) integration in intercultural or foreign environments, (4) leading across generations, and (5) fostering employee creativity and innovation (DDI 2014–15).

Structures and systems will change as organizations become more multicultural, distributed, and digitally connected. Future organizations will be built around highly empowered teams, driven by a new model of management, and led by a breed of younger, more globally diverse leaders (Deloitte 2016). In order to navigate a dynamic and connected multicultural environment, leaders will need to create an inclusive vision that can optimize knowledge and talent within the global network.

Practice: Optimal Solutions for Multicultural Collaboration

Harnessing the collective intelligence of participants in the Global Innovation and Collaboration Study (2014), future solutions were explored for facilitating multicultural collaboration during the global innovation process. The responses were gathered and grouped by themes in order to effectively identify patterns and interpret findings. The results provide insights to leadership and team development needs as well as future opportunities to improve organizational performance around the world. There are five key themes that are presented as success criteria for leading multicultural innovation and collaboration:

1. **Develop Leaders with a Multicultural Innovation Mindset**
   With organizations that have a global reach and instant access to a worldwide network, there would be an assumption that most executives and managers have extensive training and experience in cross-cultural management. However, global titles do not always equal global readiness or cultural awareness. While those who are responsible for global projects and cross-cultural teams gain substantial international experience, there is a necessity to share knowledge and experiences with all leadership levels in the organization – from managers to executives. In order to avoid cultural bias in leading global innovation, there is a need for more cultural awareness and understanding through an openness to dialogue and learning.

   Executives and leaders at all levels need to model inclusive leadership behaviors through transparency, relationship-building, and cultural empathy. Organizations that experienced challenges in multicultural collaboration often cited executives and upper level management who lacked cultural awareness and local market understanding (Jensen 2014). Thus, there is a greater need to coach and train leadership teams as well as recruit culturally diverse and experienced leaders who can effectively orchestrate global innovation across the organization.

   Leaders who can effectively facilitate multicultural collaboration create an environment where knowledge-sharing and learning can be optimized for teams across functions and geographies. They create a common goal in understanding and developing solutions that respond to local market needs. Optimizing diverse ideas and knowledge allows leaders and teams to connect with international customers and market opportunities. It requires a multicultural view with an ability to create a collective vision that shapes the direction for open collaboration, engagement, and recognition. Leadership shapes the team innovation climate where input is valued and ideas are encouraged from any part of the world. It is the opportunity to cultivate and celebrate multicultural innovation.

2. **Accelerate International Mobility and Exchange**
Although we live in a digitally connected world with 24/7 access through social media, in-person communication and live team interactions are still the most powerful and effective ways to build trust and relationships across cultures. There has been a general trend of reducing travel budgets and business trips due to the increased use of communication technologies. However, the top request from leaders working on global innovation projects is to bring everyone together in a physical location on a regular basis (Jensen 2014). Meeting in person for project launch events or planning forums allow time for team members to share, connect, and create. Budget allocation for frequent travel and interactions with local team members creates improved cultural and market understanding. Exchange visits between teams in HQ and subsidiary locations strengthen relationships as well as support of global and local strategies.

Whether scheduled for a weekend or a week, global meetings provide the opportunity to engage teams, share knowledge, and develop ideas. There is also the opportunity for international competency training where leaders and teams can integrate cross-cultural learning with the innovation project process. In addition to overseas assignments and international management training programs, some firms are experimenting with job rotations and cultural immersions where key team members can exchange locations for a specific time period in order to better understand local markets and customers. The aim is to increase cultural exposure and learning through increased international mobility that allows leaders and teams to build trust and optimize knowledge-sharing for accelerating global innovation.

3. **Create a Global Innovation Process for Cultural Contexts**
There is a growing need for a cohesive and integrated project collaboration process for global innovation, linking cross-functional and cross-cultural teams from concept to market. Most organizations are using different methodologies for the stages of front end, new product development, and go-to-market. Considerations for internationalization, local knowledge, and cultural contexts are often lacking within the project process. Moving forward, global innovation projects need to drive cross-cultural interactions in order to facilitate knowledge-sharing and collaboration amongst teams. Communication mechanisms should be integrated for

each phase in order to promote a dynamic global innovation process, moving from creation to planning to validation and execution phases.

A well-designed communications structure supports exchange and knowledge flow between teams in all international locations. It can improve engagement for cross-functional and cross-cultural teams through an integrated framework where co-creation and knowledge-sharing facilitate collaboration. Focused events and meetings with interactive agendas and real-time updates strengthen team communications. While communication technologies sustain dialogue around the world, they are still not fully optimized. Communication tools such as web conferences, video conferences, and telepresence enable visual delivery with interpersonal communication. Yet, there is still a need for more connected IT systems with shared knowledge in a single site or platform.

Organizations that are now exploring and experimenting with internal knowledge and social platforms should consider the opportunity to create a dashboard that provides a comprehensive view of the global innovation cycle and collaboration process. It should be an intelligent engine that resembles the organization's global network and facilitates conversations while assisting in finding relevant contacts and content for projects. In addition, there is the need for a built-in virtual work space and forum for seeking and generating ideas. Platforms will need to provide a way to connect local voices to global conversations, having the breadth of a global network while offering the depth of local communities. While there is still the challenge of mastering multiple languages, the evolution of language translation software may allow us to have future conversations in any language.

4. **Design Creative Work Spaces for Multicultural Collaboration**
The organizational culture and team climate serve an important role in nurturing multicultural collaboration. In order to provide team members with the opportunity to ideate and co-create, a dedicated space can be the catalyst for creativity and collaboration. A safe and open environment allows teams to foster a collaborative spirit through idea sharing and problem-solving in order to explore solutions. They can serve as work spaces for ideation, strategic planning, and project sessions where team members can share and contribute.

Visual methods and tools can engage and enable cross-cultural teams to create in collaborative spaces. Images are more powerful than the spoken word and create a common language for team members from diverse cultural backgrounds. Visuals can facilitate the expression of new ideas and co-creation of new solutions. The methods that work well for global ideation and planning sessions include graphic facilitation, idea generation and mapping, visualization, art-based exercises and proto-typing. Organizations need to capture more opportunities for spontaneous and focused creativity sessions with specific methods and tools that inspire ideas and dynamic conversations.

Organizations are designing and exploring creative spaces that foster cross-functional and cross-cultural dialogue such as creative studios and innovation hubs, from Philips InnoHub to Renault's FabLab. These spaces allow teams to connect, share, and collaborate with facilitators or team-driven activities, as well as mentoring team members across international sites. In order to succeed in developing creative spaces, there should be an off-line or physical presence as well as an online connection with geographically distributed team members. The physical space should promote in-person interactions within a dedicated environment for sessions and workshops, in addition to opportunities for networking and socializing. Moreover, the online or virtual space should build on live interactions through knowledge platforms and visual communication vehicles such as webcams and telepresence to create an interactive dialogue.

5. **Engage Teams through Local and Global Dialogue**
   In order to connect local voices to global conversations, there should be a greater emphasis on the art of dialogue. The ability to actively listen, empathize, and understand local team members and customers demands focus, observation, and attention to cross-cultural learning. Yet there is a global phenomenon that is also competing for everyone's attention - the mobile phones and communication devices that accompany team members and demand continuous screen time... In order to create more insightful dialogue, leaders and teams will have to find a blended communication mode that allows for real time interactions across geographies.

   Developing trust and motivation across cultures requires a communication process that incorporates transparency and knowledge-sharing. Leaders should consider how they frame their conversations in terms of listening and responding to the needs of their international teams and markets. This requires a structured feedback process concerning project phases. Having the opportunity to initiate ideas, receive feedback and validation creates increased motivation across cultures. Being able to share knowledge and to see the impact of contributions to innovative products and services reaffirms global team performance and recognition. Inclusive leadership behaviors can facilitate multicultural collaboration through empowerment, communication, and relationship-building. Clear roles and ownership for international teams create the opportunity to contribute valuable knowledge and insights within the global innovation network.

In order to support the above drivers and accelerate performance, incentives and evaluations will require a different set of criteria for measuring multicultural innovation and collaboration. International organizations should look at incentive structures that empower local teams through entrepreneurial initiatives and ownership within the global innovation cycle. Performance evaluations could use alternative reward and recognition systems where actual performance of the individual and team would receive equal weight to project performance. Annual reviews could include objectives for time spent on co-creation and ideation activities, performance in multicultural team collaboration and knowledge-sharing, initiatives and contributions to global and local innovation projects.

Research: Future Needs for Global and Multicultural Innovation

Global innovation inherently involves multiple countries and cultures, yet there is a lack of research and literature with theoretical and practical insights on leading multicultural innovation and collaboration. As demonstrated through qualitative research and specific company cases, this book has shown that multicultural collaboration serves a critical role in the creation and introduction of new concepts to international markets. A new research agenda should address this topic which is receiving increased attention from organizations. Currently there are few frameworks and models available that address innovation within international and multicultural contexts. There is a significant opportunity to pursue qualitative and quantitative research at organizational and managerial levels concerning the influence of multicultural collaboration on global innovation routines and practices.

There is emphasis in the literature on teamwork and collaboration within the context of new product development (NPD) and research (R&D) functions. However, there is limited attention to multicultural team collaboration during the entire global innovation cycle, from creation to planning to execution. There is research on global knowledge integration and team collaboration capabilities. However it has not explored organizational level mechanisms that influence collaboration between the global project leader and cross-cultural teams working on global innovation projects. Most of the literature separates individual level studies from organizational level studies yet multicultural collaboration often requires examination of team, organizational, and national contexts. This book attempts to fill the research gap, yet there is much to explore for multicultural collaboration by examining organizational, team, and managerial level constructs within specific contexts for global innovation.

Future research should explore multiple levels and specific contexts in order to understand multicultural collaboration. Currently, there is a greater focus on cultural value differences restricted to the individual level without consideration of other constructs including team and work unit levels (Gelfand et al. 2007). Since multicultural collaboration tends to be embedded within team, organizational, and national contexts, the analysis of macro-organizational and national contexts can enable a better understanding of how organizational factors impact the nature and use of norms, routines, and practices (Salazar and Salas 2013). There is increased interest and attention to culture as a system of meaning involving the study of social interactions, practices, and norms within a local context (Earley 2006; Pohlmann et al. 2005). Thus, researchers need to consider a holistic process in exploring cross-cultural interactions within global and local contexts.

As discussed in this book, the development of a creative space and open environment that support multicultural team collaboration serve as important factors in sustaining global innovation. Then what are the conditions that cause these spaces to accelerate or block co-creation and collaboration? In order to fully understand cross-cultural team interactions, it is necessary to examine how critical incidents emerge. Fink et al. (2006) suggest that cultural standards provide more insights since they show how critical incidents occur when choosing actions in regard to particular cultural standards and values and ignoring the other party's values and cultural standards.

Future research should employ more qualitative studies in order to acquire a better understanding of organizational, team, and individual level phenomena. Qualitative research allows for deeper insights through ethnographic methods, detailed interviews and observations. Mixed methods research can serve as a powerful methodology where quantitative study and analysis can identify phenomena combined with qualitative studies to explain gaps, patterns, and behaviors. International business and leadership contexts are increasingly critical to developing rigorous research. Attention should be paid to participant profiles where front-line leaders and managers who have global and cross-cultural team responsibilities can provide highly relevant insights.

In order to understand how to facilitate multicultural collaboration and global innovation, further research is required concerning the influence of organizational mechanisms, leadership behaviors and cross-cultural team interactions during the global innovation project. Future empirical research should consider the following topics for further exploration and validation of theory and practice:

- **Leadership and team behaviors that influence multicultural collaboration and innovation.**
- **Co-creation in front-end innovation involving multicultural teams.**
- **Global strategic planning routines between HQ and subsidiary teams.**
- **Organizational resources and routines that influence a global innovation culture and climate.**
- **Knowledge-sharing and communication practices within specific cultural contexts.**
- **Cross-cultural learning involving formal and informal methods.**
- **Multicultural innovation and global project collaboration systems, processes, and tools.**

Leading practices combined with rigorous research will accelerate the potential for developing and sustaining multicultural innovation and collaboration. The future is approaching at a rapid pace and organizations will need to harness culturally diverse talent for meeting the demands of international customers and markets. It will take the collective intelligence of a global talent pool to address new challenges and transform ideas into effective solutions around the world.

Bibliography

Deloitte Global Human Capital Trends 2016, "The New Organization: Different by Design", Deloitte University Press.

DDI Global Leadership Forecast 2014–15, DDI and The Conference Board.

Earley, P.C. "Leading Cultural Research in the Future: A Matter of Paradigm and Taste", *Journal of International Business Studies*, 37(2006): 922–931.

Fink, G. et al. "Understanding Cross-cultural Management Interaction", *International Studies of Management and Organization*, 7(2006): 38–60.

Gelfand, M.J. et al. "Cross-cultural Organizational Behavior", *Annual Review of Psychology*, 8(2007): 479–514.

The Global Innovation Index 2016, "Winning with Global Innovation", INSEAD, Johnson Cornell University, and World Intellectual Property Organization.

Jensen, Karina R. Global Innovation and Collaboration Study, 2014.
Jensen, Karina R. Local Innovation and Collaboration Study, Asia Region, 2015.
Pohlmann, Markus et al. "The Development of Innovation Systems and the Art of Innovation Management – Strategy, Control, and the Culture of Innovation", *Technology Analysis & Strategic Management*, 17(2005): 1–7.
Salazar, Maritza and Salas, Eduardo. "Reflections on Cross-cultural Collaboration Science", *Journal of Organizational Behavior*, 34(2013): 910–917.

10

Conclusion: Ready for Your World Tour?

It's time to celebrate the cultures of the world and enjoy the rich insights that each country and market bring to creativity and innovation. Listening and learning from culturally diverse perspectives nurtures an open and creative mind. Sharing country knowledge and practices enables a dynamic dialogue and engages the global network in multicultural innovation. Connecting thoughts and ideas with your team around the world transforms critical issues into solutions that empower local and global communities. Our planet is full of creative minds that are ready to share new concepts and innovative approaches shaped by diverse geographic, economic, social, and cultural contexts.

Ignoring the opportunity to learn from countries and regions across the world could be a lost opportunity to deliver value to the home market as well as international markets. Those who do not become international explorers will never have the opportunity to discover hidden gems and treasures in new lands. Denying culture and the advantage of cultural knowledge places people in an innovation vacuum of no man's land. International travel has inspired explorers and navigators for centuries, so consider launching your own innovation journey to enjoy new discoveries and insights around the globe.

Consider what kind of crew you'll need aboard the ship since you'll need team members from diverse countries, cultures and functions. Before you begin your world tour, make sure to organize a global meeting where you can create a collective vision with the international team and crew. This allows you to reflect, discuss, and design a destination that connects with everyone worldwide. Then map out your itinerary and consider which countries and regions you will visit along the way.

As a global project leader, you will need to navigate the crew through international waters, dynamic and changing markets, by serving as a knowledge facilitator and innovation orchestrator. In order to arrive at your destination, it's about harnessing the collective intelligence of your team during the journey. As discussed in Chapter 2, it starts with the ability to inspire the team through strategic co-creation in the front end, discovery and ideation through empowerment, inclusive strategic planning, a decisive and integrative process for validation. Communication and knowledge-sharing drive execution and ensure the team stays focused on global project collaboration and optimal performance.

As the global project leader and the core team move steadily forward on the journey, don't forget to ensure that everyone has specific roles that will optimize the crew's special talents and skills. As discussed in Chapter 3, strategic co-creation is most effective when local team members can serve as collaborators and intrapreneurs. Contributors may keep some team members satisfied, however the implementer roles could create less engagement, where a revolt or mutiny would result in chaos and you may never arrive at your destination.

Once you've shaped the concept and strategy, it's time to organize the roadmap. Consider key milestones and stops along the journey for ensuring that you build a concept that meets the needs of international customers awaiting the final delivery. Working with the team across geographies requires continuous dialogue at global and local levels. As noted in Chapter 4, you will need to consider diverse languages, communication styles, and ways of sharing knowledge. In order to connect with the locals, make sure you bring along the five keys to translating cultural knowledge-sharing practices.

There will be challenges and unexpected changes during the journey that may test the crew's morale. In order to keep the global team moving forward, make sure to develop trust and motivation. As presented in Chapter 5, the special practices for listening and responding during multicultural team collaboration are essentials of surviving the high seas of the global innovation project process. A balance of in person meetings with shared communication technologies and tools should keep your team happy.

As team members start enjoying the voyage and discoveries along the way, make sure to develop and sustain the global innovation culture and climate. Chapter 6 discussed values and routines for shaping a global innovation culture through cultural empathy, creativity, and collaboration. Adding to your travel kit, Chapter 7 gave some guidance on how to develop an inclusive innovation climate where teams are effectively aligned with the global innovation cycle and the project process. Keep the focus on the destination and allow your teams to thrive through an entrepreneurial spirit, global team transparency, market responsiveness, and execution efficiency.

In order to help global team leaders maintain a steady course, Chapter 8 offered a comprehensive view and reference guide to completing a successful journey, from concept to market. Leading change and organizational transformation requires specific mechanisms as discussed in previous chapters. Pay attention to the innovation strategy, the knowledge-sharing structure, the communication systems, the organizational culture and climate.

Don't forget the global innovation compass and GPS – the Multicultural Innovation Framework. Depending on your international location and multicultural innovation context, the framework can help you facilitate organizational performance and international market success. It starts by identifying the Vision necessary for leading through strategic co-creation. Next, you will need to create the Dialogue to ensure dynamic and interactive conversations, and then develop the Space necessary for an environment conducive to global project collaboration and multicultural innovation.

As you reach your final destination, it will be time to launch and deliver the treasure trove of new products and solutions that international customers have eagerly awaited. Don't forget to recognize and celebrate the end of the journey with global and local team members. A last check-in regarding lessons learned and best practices for collaboration performance is recommended before you leave ship. Once you've dis-embarked, it will be time to focus on post-launch momentum in order to accelerate marketing and sales worldwide.

For those who are ready to chart new destinations and discover new lands, Chapter 9 offered a future look and reflection on multicultural innovation and collaboration. Five organizational practices have been proposed for leaders and teams in accelerating global performance and international market success. In regard to research, there are several topics that have been proposed for further exploration and discovery of new findings while adding value to theory and practice.

Whether you are leading, managing, consulting, teaching, or researching, there are exciting opportunities for developing the art of multicultural innovation and collaboration. As demonstrated by several cases in this book – including Philips, Siemens, Adobe, Google, Lenovo, and Wipro – there are global leaders and multinational organizations that are already on their way to co-creating and orchestrating new opportunities for multicultural innovation. The first step is to listen to local market voices and to appreciate the creative potential of culturally diverse teams around the world. Wishing you a successful journey ahead!

Index

A
Adobe, 7, 8, 92, 173
Airbnb, 8, 132
Apple, 7, 35, 36
Arts-based methods, 39

B
Blue Ocean Strategy, 37
BMW, 7, 91, 92
Brainstorming, 21, 66, 71
Business feasibility, 141

C
Centralized innovation, 33, 161
Challenges in knowledge-sharing and collaboration, 9
Change levers
   Dialogue, 144, 145–146
   Space, 144, 146–147
   Vision, 144–145
Cisco, 7, 157
Climate, definition of, 120–121
Climate of innovation, 98
Collaboration capabilities, 6
Collaboration competencies, 131, 133
Collaborative economy, 149
Collaborative spaces, 51, 102
Collective intelligence, 51, 163, 169, 172
Collective vision, 103, 163, 171
Communication context, 71, 145
Communication flow, 15, 38, 149, 152
Communication, Global Innovation Cycle phases, 42
Communication, high context cultures, 59, 150
Communication, low context cultures, 59, 150
Communication styles, 14, 17, 25, 70, 149
Communication technologies, 7, 26, 30, 79, 90, 91, 106, 107, 143, 147, 149, 151, 156, 164, 165, 172
Communication tools and technologies, 106, 130

Index

Communication vehicles, 14, 30, 69, 143
Competencies, multicultural collaboration and innovation, 6, 131, 155, 162
Concept creation, 5, 19, 24, 33, 40, 45, 47, 48, 49, 80, 116, 141, 145
Conflict avoidance, 66
Contexts - international and multicultural, 167
Contexts – team, organizational and national, 119, 168
Convergent communication flows, 38
Creative spaces, 103, 148, 166, 168
Creativity, 9, 13, 14, 21, 24, 29, 39, 48, 72, 79, 86, 91, 98–102, 105, 108, 120, 121, 122, 123, 132, 143, 146, 147, 148
Cross-cultural collaboration, definition, 18
Cross-cultural communication, 59, 80
Cross-cultural learning
 formal, 154–155
 informal, 155–156
 online, 155
Cross-cultural mindset, 22, 84
Cross-cultural team interactions, 49, 155, 168
Cross-cultural teams, challenges
 conflict management, 79
 team participation, 41, 78, 146
 trust-building, 20, 21, 78, 79, 82, 144, 146, 151
Cross-functional teams, 13, 20, 36, 40, 77, 116, 119, 152, 154, 157
Cultural alignment, 69
Cultural context, 6, 23, 39, 85, 108, 150
Cultural dimension models
 Globe Study, 17
 Hall's Low Context and High Context, 17
 Hofstede's Six Dimensions, 17
 Meyer's Culture Map, 17
 Trompenaar's Seven Dimensions, 17
Cultural diversity, role of, 2, 13, 15, 98–99, 100, 102, 104, 105, 108
Cultural immersions, 119, 156, 164
Cultural interaction models
 cross-cultural Quotient, 18
 cultural intelligence, 18
 global mindset, 18
 participative competence, 18
Cultural knowledge, 45, 57, 58, 61, 101, 146, 154
Cultural sense-making, 17
Cultural standards, 101, 168
Cultural synergy, 18
Customer visits, 130

D

Decentralization, 36, 38, 45, 49
Decentralized innovation, 2, 161
Decision hierarchy, 64
Design thinking, 39, 40, 103
Developing economies, 4
Dialogue, 8, 9, 21, 22, 23, 25, 26, 28, 30, 40, 57–60, 71, 72, 78, 80, 82, 83, 84, 89, 92, 103, 106, 129, 132, 142, 144, 145, 146–149, 152, 156, 157
Dialogue, global and local, 58, 128, 145, 166–167, 172
Direct and indirect communication, 71
Discovery trips, 130
Divergent communication flows, 38
Diversity of ideas, 72
Dynamic capability, 34

E

Economic development, 4
Emerging markets, 1, 2, 4, 5, 34, 36, 91, 113
Emotional intelligence, 16

Entrepreneurial behavior, 68
Entrepreneurial initiatives, 45, 149
Essilor, 7, 89–90
Execution, 3–9, 13, 14, 16, 20–25, 27–31, 34, 36, 38, 40, 41, 44, 46, 47, 58, 59, 63, 65, 69, 79, 80, 85, 86, 87, 89, 91, 92, 100, 113, 114, 115, 116, 118, 119, 120, 121, 123, 124, 125, 130, 132, 143, 144, 145, 147, 149, 150, 152, 155, 157
Execution phase, 27–30, 36, 40, 46, 47, 79, 85, 116, 118, 119, 125, 144–145, 149
Exploitation, 37, 38, 46, 81, 113
Exploration, 8, 29, 37, 38, 40, 46, 81, 113, 114, 117, 130

F

Feedback, providing and receiving, 33, 47, 61, 68
Feedback style and delivery, 69
Formal education and training, 155
Formal learning, 153–154
Front end activities, 38
Front End Innovation (FEI), 2, 9, 20, 24, 29, 38, 40, 41, 42, 46, 49, 50, 85, 91, 114, 116, 119, 122, 129, 132, 144, 145, 154, 156, 157
Future organizations, 9, 162

G

Global communication architecture
 cross-cultural learning, 149
 global and local knowledge-sharing, 149
 social networking, 149
Global growth, 4
Global innovation, 1–3, 5–10, 12–31, 34, 38, 39, 40, 41, 45, 48, 49, 58, 60, 61, 62, 72, 79, 80, 82, 84, 85, 86, 87, 90, 91, 92, 97–108, 114–116, 119–123, 131, 132, 133, 141–148, 152–156, 161–165, 167–169, 172, 173
Global innovation climate
 entrepreneurial spirit, 121
 execution efficiency, 121
 global team transparency, 121
 market responsiveness, 121
Global Innovation and Collaboration Study, 7, 22, 45
Global innovation culture, 9, 97–108, 120–121, 146, 147, 148, 154, 156, 169, 172
Global innovation culture, capabilities, 98, 99
Global Innovation Culture, practices that nurture and sustain, 110
Global innovation culture, values
 collaboration, 9, 97, 98, 99, 100, 101, 104–106, 108, 121, 122, 146, 147, 148, 156
 creativity, 9, 98, 99, 100, 105, 108, 120, 121, 146, 148
 cultural empathy, 9, 23, 26, 99, 100–101, 105, 108, 146, 148
Global innovation cycle, 6, 7, 19, 21–24, 27, 28, 30, 31, 38, 41, 61, 80, 90, 91, 92, 107, 114, 115, 116, 132, 141, 142, 144, 146, 147, 148, 152–156
Global Innovation Cycle, five phases
 creation, 115
 execution, 115
 global launch, 115
 strategic planning, 115
 validation, 115
Global innovation process, cultural contexts, 23, 108, 163–165
Global integration, 35, 46, 140
Globalization, 4, 34, 36, 161
Global leadership, 9, 16, 17, 18–19, 132, 155, 156

Global leadership development audit, 31
Global Leadership Development Journey, 16–19, 155
Global mindset, 2, 18
Global network, 4, 23, 37, 45, 47, 58, 61, 80, 83, 92, 102, 123, 143, 149, 162, 165, 171
Global Readiness Audit, 158
Global readiness capabilities, 140
Global strategy, 25, 33, 36, 37, 41, 46, 47, 49, 78, 82, 102, 119, 143, 146
Global team performance, 6, 15, 98, 131, 144, 147, 149, 166
Global teamwork success factors, 20
Google, 7, 8, 132, 173
Go-to-Market (GTM), 9, 13, 25, 27, 39, 40, 46, 114, 116, 122, 123, 126, 132, 147, 152

H
High context culture, 59, 150
High-income economies, 161
HQ attention, 45

I
Idea hierarchy, 64, 71
Ideation(creation), 6, 9, 14, 20, 22, 33, 40, 41, 46, 68, 71, 72, 86, 88, 102, 105, 106, 107, 114, 115, 119, 130, 139, 143, 144, 147, 152
Inclusive dialogue, 84, 146, 156
Inclusive innovation climate, 113–133, 172
Inclusive leadership, 23, 25, 26, 29–31, 144, 148, 157
Inclusive leadership behaviors, 26, 163, 166
Incremental innovation, 37, 38, 39, 91, 116
Individual level and organizational level studies, 168
Informal and formal interactions, 61
Informal learning, 153–154, 155, 156
Initiative-taking, 71, 145
Innovation orchestrator, 6, 9, 157, 172
Innovation strategy, 14, 46, 49, 92, 142
Innovative thinking, 99, 102, 105, 108, 146
Integration-Responsiveness framework, 35
Intel, 7
Internationalization, 16, 35, 36, 116, 118, 119, 125, 147
International mobility, 92, 104, 106, 146, 155, 164
International strategy, 1, 35, 82, 129
Interpretations, 60
Intracultural competence model, 18

K
Knowledge-based concept of cross-cultural management, 61
Knowledge facilitator, 9, 20, 28, 30, 87, 147–148, 172
Knowledge flow, headquarters and subsidiaries, 44
Knowledge platform, 68, 80, 90, 107, 141, 143, 147, 149, 152, 156
Knowledge-sharing, 1, 2, 3, 6, 9, 14, 15, 18–21, 23, 25, 26, 28–30, 38, 40–42, 45–47, 50, 58–68, 70, 71, 79–82, 85–90, 92, 98–100, 103–108, 114, 115, 119–122, 124, 125, 129, 131, 132, 140, 141, 143–147, 150, 152, 153, 156, 157
Knowledge-sharing challenges
communication, 21, 30, 38, 40, 47, 58, 60, 71, 79, 80, 81, 87, 90, 92, 103, 119
feedback, 89, 119

HQ and subsidiary roles, 79, 80
local and global visibility, 79, 80
local market intelligence, 21, 23, 41, 79, 129
Knowledge-sharing, cultural implications, 145
Knowledge-sharing, definition of, 61
Knowledge-sharing forums and tools, 136
Knowledge-sharing platform, 90, 153
Knowledge-sharing process, cultural differences
  initiative, 62
  openness, 62
  power, 62
  response, 62
  structure, 62
Knowledge-sharing structure, 14, 45, 90, 107, 143

L

Language, 13, 14, 21, 25, 28, 34, 36, 39, 58, 59, 60, 66, 68, 80, 114, 117, 118, 148, 150, 151, 155
Leaders, facilitators and orchestrators, 9, 30–31
Leadership
  definition of, 16
  role of power, 17, 30, 145
Leadership and team behaviors, 122, 169
Leadership behaviors
  communicative, 23, 28
  directive, 23, 27
  empowering, 23, 24
  inclusive, 23, 25, 26
Leadership behaviors in Asia, 29–30
Leadership behaviors within the global innovation cycle
  creation phase, 23, 24, 29, 30, 31, 119
  execution phase, 27, 28, 145

planning phase, 23, 25, 26, 31
validation phase, 23, 27, 30, 40, 41, 47, 119, 144
Leadership theories
  behavioral, 16, 18
  contingency, 16
  transactional, 16
  transformational, 16, 17
Lenovo, 7, 8, 104, 173
Local customer references, 125–126
Local immersions, 130
Local Innovation and Collaboration Study, 8
Localization, 14, 16, 36, 114, 116, 118, 120, 124, 125, 145, 147
Local language context, 60
Local management views, Asia, 29, 82, 139
Local market insights, 125
Local market intelligence, 21, 23, 41, 79, 129, 130, 131
Local market knowledge, 9, 24, 38, 41, 46, 47, 48, 49, 63, 79, 82, 88, 99, 129, 145
Local market knowledge, formal and informal methods, 129
Local responsiveness, 35, 46, 99, 100
Low context culture, 59, 150
Low income economies, 5, 161

M

Managerial leadership, 16
Marketing and sales, 34, 36, 116, 118, 124–126, 132, 140
Market opportunity, 38, 141
Mature markets, 4, 5, 35
Mazda, 7, 91, 92
Middle-income economies, 161
Mixed methods research, 168
Motivation and multicultural team collaboration
  engagement, 85

Index

Motivation and multicultural team collaboration (cont.)
  listening, 85
  local connection, 85
  recognition, 85
  responsiveness, 85
Multicultural innovation and collaboration, 6–10, 18, 19, 21, 61, 82, 105, 157, 161–169, 173
Multicultural Innovation Framework, 8, 18, 142, 144, 173
Multicultural innovation mindset, 156, 163–164
Multidomestic or multilocal strategy, 35

N

Nambisan and Sawhney, 4
Network-centric innovation, 4
New product development (NPD), 39, 40, 114, 115, 116, 122, 123, 132, 147
New product introduction, 33, 39, 91, 123
Nokia, 7

O

Open innovation, 5
Orchestration of innovation process, 157
Orchestration of organizational mechanisms, 142
Organizational culture, 9, 14, 48, 62, 77, 98, 104, 105, 108, 121–123, 132, 143, 144, 150, 157
Organizational mechanisms, 14, 15, 141, 142, 157
Organizational performance, success criteria, 9, 15, 16, 132, 142, 163, 173
Organizational resources and routines, 4, 29, 50, 141, 169
Organizational routines
  Cross-cultural team interactions, 105
  Inclusive team leadership, 105
  project collaboration and communication tools, 106
  project process, 105
  shared workspace, 105
Organizational strengths, 99–100
Organizational transformation, 16, 17, 173

P

Performance, team and project, 42, 131, 132–133
Philips, 7, 8, 166, 173
Planning and execution phases, 40, 79, 85, 116, 118, 149
Planning and execution phases, knowledge requirements, 40, 79, 118
Planning phase, 23, 25, 26, 29–31, 36, 40, 41, 44, 47, 48, 49, 79, 87, 118, 119, 123, 124, 132, 145
Power, 4, 16, 17, 62, 64, 65, 66, 71, 85, 145
Power structures, 85
Project collaboration, 9, 14, 15, 17, 18, 20, 22, 23, 25, 27, 28, 30, 31, 41, 45, 46, 50, 59, 60, 61, 67, 67, 71, 81, 84, 85, 86, 92, 106, 107, 114, 115, 119, 125, 132, 141, 143, 145, 148–152, 154–156
Project collaboration process, 9, 23, 30, 41, 46, 60, 61, 107, 119, 132, 145, 149, 150
Project collaboration roles
  collaborator, 45
  contributor, 45

implementer, 45
intrapreneur, 45
Project execution, 30, 58, 59, 69, 150
Project management process, 27, 39, 117
Project performance, 42, 59, 98, 131, 133, 145, 147
Project performance measures, 42, 131

R

Radical innovation, 36, 37
Relationship-building, 2, 20, 21, 22, 23, 25, 26, 29, 30, 59, 82, 84, 92, 107, 132, 144, 147, 150, 151
Research, global and multicultural innovation, 9–10, 167–169
Reverse innovation, 5
Risk-taking, 48, 80, 100, 121, 123
Roadmap, 9, 13, 117, 118, 172
Roles of headquarters and subsidiaries, 38

S

Sales readiness, 116, 120, 126, 132, 147
Samsung, 7, 35, 36
Saving face, 66, 71
Self-service knowledge, 68
Sense-making, 17, 39
Siemens, 7, 8, 108, 173
Siemens Convergence Creators, 108
Smart phone market, 35
Social innovation, 5
Social making, 39
Social networking
  formal, 150
  informal, 151
  virtual, 151–152
Space, 8, 9, 20, 21, 22, 24, 26, 39, 80, 81, 90, 98, 101, 103, 105–108, 122, 132, 142, 144, 146–148, 156, 157

Strategic co-creation, 9, 25, 26, 39, 48, 49, 92, 119, 143, 144, 145, 148, 156, 172, 173
Strategic Co-creation Audit, 51
Strategic dimensions
  content, 34
  context, 34
  process, 34
Strategic planning, 1, 6, 20, 21, 23, 25, 26, 29–31, 33, 36, 40, 41, 44, 47–49, 68, 82, 115–117, 120, 124, 129, 139, 145, 147, 150, 157
Strategic planning phase, 23, 26, 29, 30, 36, 40, 44, 47, 49, 145
Strategy-making, 2, 34, 38, 39, 40, 49
Subsidiary roles and knowledge flows, 37
Subsidiary roles and strategic importance, 36–37

T

Team climate, 9, 14, 115, 121, 132, 143, 144, 146, 147, 154, 157
Team innovation climate, 113, 124, 147, 148–149
Team performance, 6, 15, 23, 31, 42, 98, 131, 132, 144, 145, 147, 149
Transnational strategy, 35, 36
Transparency, 6, 47, 67, 81, 83, 84, 85, 86, 88, 92, 97, 99, 100, 103–105, 108, 121, 122, 124, 132, 140, 143, 146, 147–149, 152
Trust and motivation, 69, 87, 88, 92, 145, 146, 166, 172
Trust-building
  listening, 23, 59, 78, 82, 85, 87, 88, 89, 92, 146
  open communication, 79, 82, 84, 88, 92
  project contribution, 82, 84
  response, 69, 82, 84, 89, 145

Trust-building (cont.)
  relationship-building, 20, 21–23, 26, 30, 59, 82, 84, 92, 108, 132, 144, 150, 151
Twitter, 132

V
Validation, 6, 14, 22, 23, 27, 28, 29–31, 36, 38, 40, 41, 46–47, 68, 71, 86, 88, 92, 106, 115–120, 125, 129, 130, 142, 144, 145, 146, 152
Value chain, 3, 31, 34, 35, 141
Vision, 6, 8, 9, 16, 20, 22, 23, 24, 27, 30, 39, 40, 48, 59, 61, 91, 98, 103, 142, 144, 148, 156, 157
Visual communication, 102–103, 146, 152
Visualization, 103, 166
Visual methods and tools, 107, 165

W
Wipro, 8, 129, 173
Work spaces, 90, 106, 147